Springer Series in Synergetics   Editor: Hermann Haken

Synergetics, an interdisciplinary field of research, is concerned with the cooperation of individual parts of a system that produces macroscopic spatial, temporal or functional structures. It deals with deterministic as well as stochastic processes.

L.A. Blumenfeld

Physics of Bioenergetic Processes

With 30 Figures

Springer-Verlag
Berlin Heidelberg New York Tokyo 1983

Professor Lev Alexandrovitsch Blumenfeld

Institute of Chemical Physics, Academy of Sciences, Kosygin Street 4, V-334
Moscow, GSP-1, 117977, USSR

Volume Editor:

Professor Dr. Dr. h. c. Hermann Haken

Institut für Theoretische Physik der Universität Stuttgart, Pfaffenwaldring 57/IV
D-7000 Stuttgart 80, Fed. Rep. of Germany

ISBN-13:978-3-642-68527-9   e-ISBN-13:978-3-642-68525-5
DOI: 10.1007/978-3-642-68525-5

Library of Congress Cataloging in Publication Data. Bluimenfel'd, L. A. (Lev Aleksandrovich) Physics of bioenergetic processes. (Springer series in synergetics ; v. 16) Bibliography: p. Includes index. 1. Bioenergetics. 2.
Biophysics. I. Title. II. Series. [DNLM: 1. Energy metabolism. 2. Biophysics. QU 125 B649p]   QH510.B54
1983   574.19'121   83-4245

Foreword

According to its definition, synergetics is concerned with the cooperation of individual parts of a system that produces macroscopic temporal, spatial or functional structures. A good deal of the volumes published within this series dealt with the formation of truly macroscopic structures which we can see with our eyes. A common scheme could be developed to understand the formation of many patterns through self-organization. In particular, we have to use concepts which go beyond conventional thermodynamics. New ideas became crucial. We have to study kinetic processes, and often few highly excited degrees of freedom play the decisive role in the evolution of structures. Over the past years it has turned out that quite similar lines of approach apply to a world which at first sight would be classified as "microscopic". That world consists of processes in which biomolecules are involved. An important example for the problems occurring there is provided by Manfred Eigen's theory of evolution of life at the molecular level (cf. his contribution to Volume 17 of this series). Another important example has been provided by Blumenfeld's book on problems of biological physics (Vol.7 of this series). There it was proposed to treat biological molecules as machines which, in a certain sense, work through "macroscopic" degrees of freedom. The inadequacy of concepts of equilibrium thermodynamics or irreversible thermodynamics was pointed out and the processes were studied in detail through kinetic models. This line of approach is again taken up in the present book by L.A. Blumenfeld, which deals with various aspects of bioenergetic processes such as muscle contraction, active transport of ions, substrate and membrane phosphorylations. He treats proteins as "molecular machines" and stresses the importance of nonequilibrium states of proteins. The important feature of Blumenfeld's book consists in introducing physical concepts into the bioenergetic processes of molecules whereby he sheds new light on enzymatic catalysis.

In view of the extremely complex processes which go on in living matter some of Blumenfeld's ideas might need further elaboration. But his approach opens new vistas in the treatment of these processes and a study of his book seems to be a must for all those who try to understand this kind of processes. For these reasons I am very happy that we can include Blumenfeld's book in the Springer Series in Synergetics.

Stuttgart, June 1983   *Hermann Haken*

Acknowledgement

Over the recent years the subject matter of this monograph has been repeatedly discussed with my coworkers and colleagues. These discussions were extremely useful and did help me to remove many errors (not all of them, I'm sure) and to formulate more clearly the critical and the positive statements.

I wish to express my deep gratitude to B. Cartling, A. Ehrenberg (Arrhenius Laboratory, Stockholm University), D.Sh. Burbajev, M.G. Goldfeld, R.M. Davydov, A.F. Vanin (Institute of Chemical Physics, USSR Academy of Sciences), L.V. Jakovenko, V.I. Pasechnik, A.N. Tikhonov, V.A. Tverdislov (School of Physics, Moscow University), L.I. Boguslavsky, Yu.A. Chismadzhev (Institute of Electrochemistry, USSR Academy of Sciences), D.S. Chermavsky (Physical Institute, USSR Academy of Sciences).

I also thank N.P. Kirpichnikova and N.V. Vojevodskaja for their help in the preparation of figures.

Moscow, March 1983   L.A. Blumenfeld

Contents

1. Introduction

Achievements of biological chemistry in recent decades are striking. This is true
not only for current popular molecular genetics but for all branches of science
dealing with chemical compounds of living matter and their transformations. In many
laboratories throughout the world highly qualified scientists use first-class physical instruments and sophisticated chemical and biological techniques to isolate,
purify, and characterize low- and high-molecular compounds responsible for the
functioning of the chemical machinery of cell and tissues, in order to determine
the chemical mechanisms and the kinetic and thermodynamic parameters of biochemical
processes.

All this also holds true for that branch of biochemistry (or biophysics, biology,
physicochemical biology, biophysical chemistry, molecular biology—the number of
different names denoting one and the same thing grows exponentially with the number
of scientists and scientific publications), which is the subject matter of this book.
The volume of material bearing relation to bioenergetic processes is becoming truly
immense.

At the same time reading of original articles, reviews, and monographs dealing
with the mechanisms of muscle contraction, active transport of ions, or membrane
phosphorylation gives one not only an almost aesthetic pleasure (always produced by
descriptions of beautiful experiments) but a deep feeling of dissatisfaction as
well. This situation was already exhaustively described in 1947 by SZENT-GYORGUI
[1.1] who said that as we knew more and more about muscle contraction we understand
less and less, and that we should soon know everything and understand nothing. The
situation has not improved up to now. This becomes clear, e.g., if we consider the
processes of membrane phosphorylation, i.e., the most important processes of transduction and accumulation of light quanta energy in plants and photosynthetic bacteria, and of food oxidation energy in the mitochondria of animals and microbes.

Unbiased scrutiny of existing theories reveals their intrinsic contradictions
and the impossibility to conform them to experimental data. The best illustration of
this can probably be found in the presently most popular chemiosmotic concept of
membrane phosphorylation mechanism proposed in 1961 by MITCHELL [1.2]. This concept
was not readily accepted by scientific public opinion.[1] (Footnote 1 see next page.)
At first the chemiosmotic approach seemed to be too revolutionary and unwanted as

compared with the orthodox postulates of chemical concept popular in those days. Experimental data, which could be explained by Mitchell's hypothesis, were, however, gradually accumulated, some predictions of this hypothesis were verified in several laboratories, and in the 1970s the chemiosmotic concept became quite respectable and was universally accepted.

I fully understand that Mitchell's concept was extremely helpful in stimulating a great number of beautiful experiments which essentially enlarged our knowledge of the membrane transfer of protons and other particles, i.e., of the processes whose importance in the regulation of intracellular electron transport and energy transduction is beyond doubt. I have always thought, however, that the physical fundamentals of this concept (i.e., of the postulate that the energy source for membrane phosphorylation is the transmembrane electrochemical potential gradient) are rather doubtful. Detailed examination of the chemiosmotic concept, and its relation to experimental data can, at the same time, help us provide an answer to a more general and probably more important question: what is the reason for theoretical flimsiness of modern biochemistry when its experimental and practical achievements are so remarkable. This reason may be formulated as follows: twentieth century biochemistry uses for theoretical description of biochemical processes nineteenth century physical chemistry, i.e., the concepts developed to describe the behavior of low-molecular compounds in gaseous phase and dilute solutions. In thermodynamics this means the unreserved use of the Van't Hoff equation, and in kinetics of the Arrhenius equation (or, which is the same, of the activated complex theory). I am speaking, of course, not only about the formal use of corresponding mathematical expressions in calculating the thermodynamic and kinetic parameters of chemical reactions, but about the acceptance of postulates underlying these expressions, and the acceptance of the physical models of processes for which these expressions have been derived. One of the aims of the present monograph is to prove the inadequacy of this physical model for the majority of biochemical processes, particularly for the processes of energy transduction in biological systems.

Because of this inadequacy many experimental facts described in scientific publications are not facts in the true sense of the word. For example, the statement "the equilibrium constant of this reaction equals to..." represents for many enzymatic reaction not an experimental fact but its interpretation. The experimental facts in this case are the measured values of reagent concentrations constant with time in the reaction mixture. The statement concerning the equilibrium constant is in this case an interpretation based on the assumption that the mass action law holds true for this reaction. This assumption for a process involving ordered macro-

---

1 All scientists that have worked long enough know that in science public opinion probably plays a more important role than in other human undertakings. Respectability of research directions and points of view often determine for many years the progress of a branch of science and the status of a scientist.

molecular structures cannot be considered a priori as true, and requires special discussion.

This monograph is to a certain extent the continuation of the book *Problems of Biological Physics* [1.3] the first edition of which was published in the USSR in 1974 [1.4]. The theoretical basis in the present book, as it was in [1.3], is the concept that the physical mechanisms of processes involving high-ordered macromolecules require, as a rule, not a statistical, but a mechanical approach. This is primarily true for bioenergetic processes briefly discussed in the last chapter of [1.3]. These processes are actualized by means of elementary chemico-mechanical transducers — the machines of molecular dimensions.

In several laboratories of different countries experimental and theoretical studies directly or indirectly based on these ideas have been carried out in recent years. My co-workers in the Laboratory of the Physical Chemistry of Biopolymers (Institute of Chemical Physics, USSR Academy of Sciences) and in the Biophysical Department (School of Physics, Moscow University) have obtained new results, and bioenergetic processes can now be discussed much more comprehensively than they were in [1.3]. All the aforesaid can be considered to justify this book.

2. Phenomenology of Bioenergetic Processes

In this chapter the basic data concerning the most important bioenergetic processes will be given. These data are necessary to grasp the essence of solved and unsolved problems arising when we consider the relevant physical mechanisms. During recent decades scientists of different specialities in many laboratories throughout the world have obtained numerous facts concerning the structures of intracellular energy-transducing systems, and the biochemical mechanisms of corresponding reactions. It is impossible to touch even briefly upon all these highly interesting questions in such a short book. Therefore in the first chapter I shall describe only the strictly established and generally accepted facts without the knowledge of which the reader will not be able to understand the solutions let alone the formulations of corresponding physical problems. More detailed descriptions of experimental data can be found in the list of references.

2.1 Muscle Contraction

The approach taken to bioenergetic problems is based on a postulate according to which all the bioenergetic processes are actualized by means of elementary chemico-mechanical transducers, that is the molecular machines capable of exciting specific degrees of freedom and realizing the coherent states of mechanical motion by utilizing the local statistical acts of chemical transformations [2.1-3]. We shall see that mechanisms of this type are able to provide for various bioenergetic processes: overcoming an activation barrier in enzymatic catalysis, transferring ions across membranes against their electrochemical potential gradients, synthesizing thermodynamically unfavorable compounds by means of certain energy-donating reactions. It is natural to begin with the processes for which the excitation of specific degrees of freedom and mechanical motion directly represent the purpose of the functioning of elementary chemico-mechanical transducers, i.e., with the processes of biological motility.

Biological motility is a traditional field of biophysical research. All living things move (although not everything that moves is living). Macroscopic (mechanical) motion is the first phenomenon confronting a scientist studying biological objects at different levels of organization. Some examples are the continuous motion of protoplasm, mechanical motion of cell components in the course of cell life cycle, the motion of spermatozoa, various kinds of muscle activity, the relative motion of ribosomes and messenger RNA molecules during protein biosynthesis. The structure and phenomenology of the functioning of striated muscle are known quite well. Since the physico-chemical principles of the contraction of all biological systems are rather alike, it is therefore appropriate, to describe here the contraction process of vertebrate striated muscle.

An elementary part of a muscle contractile fiber is a polynuclear cell with thickness of 20-100 µm and with a length of up to a few cm. The cell is filled with sarcoplasmatic reticulum and subcellular particles (nuclei, mitochondria) but a substantial part of its volume is occupied by myofibriles — elongated contractile structures with a diameter of 1-2 µm (one fiber contains $\sim 10^3$ myofibrils). Myofibrils contain two types of filaments, thick and thin ones, packed in comparatively small repeating units, the so-called sarcomeres, with a length (for a resting muscle) of ~ 2.2 µm. Sarcomeres are separated by protein Z-discs, with which thin filaments are connected. Thin filaments are mainly composed of actin, but also include Ca^{2+}-sensitive regulatory complexes of the proteins tropomyosin and troponin. An actin filament is a fibrillar polymer of monomeric actin globules (G-actin, molecular weight ~ 46000 D, diameter ~ 5.5 nm) packed in a double helix with an axial period of ~ 71 nm. Thick filaments are composed of myosin molecules (about 350 protein molecules per one thick filament). The elongated myosin molecule, molecular weight (MW):~ 470000 D, has a length of ~ 150 nm and a diameter (in the core region) of ~ 2 nm. The thickened end of a myosin molecule (the so-called head) has a length of ~ 20 nm and a diameter of ~ 4 nm. Each sarcomere contains $\sim 10^6$ thick and thin filaments. The structure of resting muscle at different organization levels is shown in Fig.2.1. In thick filaments the heads of myosin molecules form periodically localized projections, the miosin cross-bridges. The axial distance between identically positioned bridges equals ~ 43 nm.

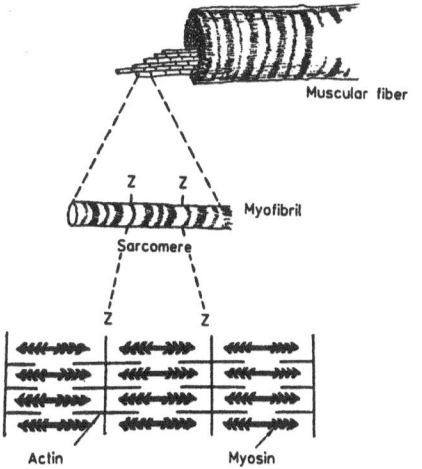

Muscular fiber

Z   Z   Myofibril

Sarcomere

Z   Z

Actin   Myosin

Fig.2.1. The structure of a resting muscle at different organization levels

5

In a resting muscle the myosin and actin filaments do not interact. After muscle activation the myosin bridges bind to the actin monomers of thin filaments. The acto-myosin complex plays a role of a chemicomechanical energy transducer. During muscle contraction the length of thin and thick filaments does not change: they slide relative to one another. This theory of muscle contraction, "the sliding filament model" [2.4], is now generally accepted. Filament structures do not change during contraction. Only the relative positions of bridges are changed, since bridges represent the sole movable components of the whole structure. Not discussed here are Ca^{2+}-dependent displacements of the troponin—tropomyosin complex relative to the actin filament, which have a regulatory function making actin centers available to myosin bridges [2.5].

The sequence of processes occurring during muscle contraction can be visualized as follows. The binding of Ca^{2+} ions by troponin-tropomyosin complex leads to conformational changes in protein thin filaments. These conformational changes result in the "opening" of monomeric actin "beads" and, consequently, in the attachment of myosin bridges to corresponding (i.e., those situated at this moment of time opposite them) centers on thin filaments. In its attached state every bridge exerts a pushing force which pushes the corresponding thin filament to the sarcomere center. This leads to the sliding of thin filaments relative to thick ones and, consequently, to the diminishing of sarcomere length. The pushing force value depends on the bridge displacement (i.e., on the bridge deformation relative to its equilibrium position in a free state), and if a displacement is large enough the pushing force changes its sign, i.e., becomes a hindering force. The sliding of filaments leads not only to a change in the pushing force but also to a change in the binding constant of a myosin bridge with actin centers, so the probability of dissociation increases. The bridge passes from the hindering to the free state, relaxes to equilibrium conformation relative to the myosin framework and is now capable of binding again the next free actin center. In the region of thin and thick filaments, overlapping myosin bridges function independently—the forces exerted by them are additive.

An elementary mechanical cycle for one bridge can be formally represented by the following scheme [2.6]:

$$a \longrightarrow b \qquad . \tag{2.1}$$
$$\searrow c \swarrow$$

Here a, b, c are states of free bridge, attached pushing bridge, and attached hindering bridge, respectively. The rate of $b \rightarrow c$ transitions depends on the filament-sliding velocity. Each myosin bridge is thus largely an independent mechanical arrangement able to perform mechanical work. This molecular machine functions cyclically. The work is performed at the $b \rightarrow c$ stage of (2.1). The energy source of the cyclic process (2.1) is enzymatic ATP hydrolysis. The discovery of actomyosin ATPase activity by ENGELHARDT and LYUBIMOVA in 1939 [2.7] was the first step in the analysis of chemicomechanical energy transducers of living matter. Today, the biochemistry of myosin and actomyosin ATP hydrolysis by isolated proteins in solution is known rather well.

Let us first consider the data concerning the catalytic activity of myosin. As a matter of fact all research has been carried out under physiological conditions (pH, ionic strength) at which myosin is insoluble. Therefore, as a rule, the biochemical activity of heavy meromyosin or of subfragment -1 (S-1) was studied (heavy meromyosin is soluble fragment of myosin containing catalytic center; S-1 is catalytically active subunits of heavy meromyosin whose molecular weight is about 1/3 of that of the latter). It can be assumed that enzymatic properties of myosin, heavy meromyosin and S-1 are alike.

It has been established that during ATP hydrolysis protein conformation undergoes marked changes, and the protein intermediate state differs from the free myosin state and the equilibrium complex state of myosin with reaction products (ADP and P_i) as well [2.8-13]. Heat liberation during ATP hydrolysis by myosin in solution is realized not at the time moment of ATP ester-phosphate bond splitting but after 15-20 s [2.14]. One can conclude that after hydrolysis the energy is stored for some time in the conformationally nonequilibrium state of the protein.

Detailed investigations carried out in different laboratories [2.15-20] have resulted in a currently generally accepted scheme of ATP hydrolysis by myosin. This scheme includes ordinary chemical reactions and the stages of protein conformational changes:

$$M + ATP \underset{k_{-1}}{\overset{k_{+1}}{\rightleftarrows}} M \cdot ATP \underset{k_{-2}}{\overset{k_{+2}}{\rightleftarrows}} M^* \cdot ATP \underset{k_{-3}}{\overset{k_{+3}}{\rightleftarrows}} M^{**} \cdot ATP \underset{k_{-4}}{\overset{k_{+4}}{\rightleftarrows}} M^* \cdot ADP \cdot P_i$$

$$\underset{k_{-5}}{\overset{k_{+5}}{\rightleftarrows}} M^* \cdot ADP + P_i \underset{k_{-6}}{\overset{k_{+6}}{\rightleftarrows}} M \cdot ADP + P_i \underset{k_{-7}}{\overset{k_{+7}}{\rightleftarrows}} M + ADP + P_i \quad . \tag{2.2}$$

Here superscripts $*$ and $**$ designate different conformationally changed protein states. At room temperatures (21-24°C), pH 8.0 and ionic strength 0.1, the following values of equilibrium constants (K_i) and rate constants (k_i) for individual stages of (2.2) have been obtained:

$$K_1 = 4.5 \times 10^3 M^{-1}; \quad K_2 > 2 \ 10^4; \quad k_{+2} = 400 \ s^{-1}; \quad k_{-2} < 0.02 \ s^{-1};$$

$$K_3 = 9 \times k_{+3} = 160 \ s^{-1}; \ k_{-3} = 18 \ s^{-1}; \quad K_4 < 2 \times 10^7; \quad k_{+4} = 0.06 \ s^{-1};$$

$$k_{-4} > 3 \times 10^{-9} \ s^{-1}; \quad K_5 > 1.5 \times 10^{-3} \ M; \ K_6 = 3.5 \times 10^{-3}; \ K_7 = 2.7 \times 10^{-4} \ M \quad .$$

Equilibrium constant for the overall reaction (ATP $\rightleftarrows$ ADP + P_i) under these conditions is $K = K_1 K_2 ... K_7 = 1.7 \times 10^7$ M [2.21].

These values show that in the course of ATP hydrolysis the free energy of the system decreases mainly at stages 2 and 4 connected with conformational changes of macromolecular enzyme—substrate and enzyme—product complexes. The slowest stage is not the act of ATP hydrolysis (stage 3) but the conformational change of enzyme—product complex following stage 3 and preceding the liberation of products. Some constants (K_2, k_{-2}, K_4, k_{-4}, K_5) have not been determined precisely. The "true" values (i.e., the values satisfying (2.2) and the overall reaction equilibrium constant K) cannot, however, differ greatly from the upper or lowest limiting values given above. Some of these constants were measured for ATP analogs and the "true" values almost coinciding with the above limiting values were obtained. For instance, k_{-2} value for ATP (β, γ-NH) (S'-adenylimidodiphosphate) is exactly $0.02 \ s^{-1}$ [2.19]. We shall return later to the discussion of the reality of scheme (2.2) and the values of thermodynamic and kinetic constants of its individual stages.

ATP hydrolysis by myosin is a rather slow process. In muscle fibers enzymatic hydrolysis and energy liberation proceed by 2-3 orders of magnitude faster, which is causally related to the periodic formation and dissociation of actomyosin complexes, i.e., to the attachment and dissociation of myosin bridges.

Deshchervsky's model of an elementary mechanical cycle of striated muscle contraction described above does not refer mechanical stages to biochemical ones. This can be done, however, if we assume that biochemical mechanism of enzymatic ATP hydrolysis by ordered actomyosin systems in a muscle resembles that of ATP hydrolysis by soluble actomyosin complexes studied experimentally.

The currently generally accepted cyclic mechanism of actomyosin ATPase was independently proposed by LYMN and TAYLOR [2.22], and by BUKATINA and DESHCHEREVSKY [2.23] (see also [2.3,24]). The most important and apparently surprising experimental fact is the observation that although actin accelerates ATP hydrolysis by myosin about 200 times [2.25], the ATP addition to actomyosin leads to a rise in the actomyosin effective dissociation constant. It means that during ATP hydrolysis actomyosin disappears, and the concentration of a less effective catalyst, myosin, increases. Analysis carried out in [2.22,23] has, however, shown that actin accelerates the stages of reaction product dissociation, and not the stage of ATP hydrolysis. A rather simplified scheme of ATP hydrolysis by actomyosin according to the authors cited above can be presented as follows:

$$
\begin{array}{ccccccc}
 & 1 & & 2 & & 3 & \\
S+M \rightleftharpoons & M \cdot S \rightleftharpoons & M \cdot Pr \rightleftharpoons & M+Pr \\
\end{array}
$$

$$(2.3)$$

Here S is ATP and Pr is ADP + P_i. The rate constants of stages 1-10 have such values that at a normal physiological substrate and not extremely low actin concentrations the following sequence of stages is realized: 9,8,2,6,5. It means that (2.3) is transformed into

$$S + A \cdot M \xrightarrow{10^6} A \cdot M \cdot S \underset{-A}{\xrightarrow{10^3}} M \cdot S \xrightarrow{150} M \cdot Pr \underset{+A}{\xrightarrow{3 \cdot 10^5}} A \cdot M \cdot Pr \xrightarrow{20} A \cdot M + Pr \quad . \quad (2.4)$$

The values of corresponding rate constants (numbers above the arrows) are given in $M^{-1}s^{-1}$ units (for bimolecular stages) and s^{-1} units (for monomolecular stages). ATP addition to the actomyosin complex results in a fast dissociation of the latter. ATP hydrolysis is realized mainly in the ATP-myosin complex. The ADP and P_i dissociation from myosin, however, proceeds so slowly that the reaction goes predominantly along the following path: actin attachment to the MPr complex and product dissociation from actomyosin.

The actomyosin complex thus undergoes a full dissociation and formation cycle during which one ATP molecule is hydrolyzed. DESHCHEREVSKY [2.3] correlates this biochemical cycle with an elementary mechanical cycle (2.1). The free bridge (a) represents a mixture of biochemical MS and MPr states; the pushing state (b), an AMPr complex at earlier stages of its conformational relaxation; and the hindering state (c), a mixture of AM, AMS and AMPr at the last stages of its conformational relaxation.

Deshcherevsky has also proposed and analyzed a kinetic model of muscle protein functioning, in which the regulation of the process by the proteins of the troponin-tropomyosin complexes and Ca^{2+} ions has been accounted for [2.3].

Schemes (2.2) and (2.3) require discussing. The constants were estimated assuming that at any stage of hydrolysis the intermediates have had enough time to reach equilibrium states. Prior to reaching complete thermodynamic equilibrium in the system ATP, ADP and P_i, only slowly changing net chemical composition of the reaction mixture remains out of equilibrium. This composition is regarded as a parameter. At every given moment of time (i.e., at a given parameter value) the system is assumed to be at équilibrium. It means, in particular, that k_i values do not depend on the concentrations of the final and initial products and that during a reverse process, i.e., ATP synthesis from ADP and P_i, the enzyme molecule passes through the same states as during the direct process but in a backward sequence.

The validity of these assumptions is rather doubtful. The fact that such schemes as (2.2,3) allow one to describe without contradiction a limited set of experimental kinetic data cannot be considered as proof of these schemes and their underlying postulates. Every scientist working in the field of chemical kinetics knows that kinetics data are not able to prove a proposed mechanism; they can at best refute it. In the case of a process, some stages of which are certainly accompanied by considerable conformational changes of protein macromolecules, the assumption that all stages during this process have enough time to reach equilibrium is very improbable.

Extremely important is, therefore, the work [2.26] in which ATP synthesis from ADP and P_i was studied for the same myosin subfragment S1 with which the thermodynamic and kinetic constants of ATP hydrolysis in the above-mentioned works were determined. It was shown that ADP and P_i addition to S1 leads to the formation of surprisingly large quantities of ATP [in the experiments the total ATP: ATP + M × ATP + M^{X}ATP was measured, see scheme (2.2)].

The authors op cit. succeeded in obtaining the exact values of equilibrium constants K_4 and K_5, for which the analysis of the direct reaction could give only the highest and the lowest limiting values, respectively [2.18,19]. The K_5 value $(7.3 \times 10^{-3}$ M) lies close to the lowest limiting value estimated in [2.19] $(K_5 > 1.5 \times 10^{-3}$ M), the K_4 value obtained was, however, surprisingly low, $-K_4 = 15.6$, i.e., by ∼6 orders of magnitude lower than the upper limit estimated in [2.19]. Summing up the data of hydrolysis [2.15-20] and synthesis [2.26] and the thermodynamic characteristics of the overall process [2.21], one can estimate the values of all the thermodynamic and the majority of kinetic constants in (2.2):

$$K_1 = 4.5 \times 10^3 M^{-1}; \quad K_2 = 2 \times 10^9; \quad k_{+2} = 400\ s^{-1}; \quad k_{-2} = 2 \times 10^{-7}\ s^{-1}; \quad K_3 = 9;$$

$$k_{+3} = 160\ s^{-1}; \quad k_{-3} = 18\ s^{-1}; \quad K_4 = 15.6; \quad k_{+4} = 0.06\ s^{-1}; \quad k_{-4} = 4 \times 10^{-3} s^{-1};$$

$$K_5 = 7.3 \times 10^{-3}; \quad K_6 = 3.5 \times 10^{-3}M; \quad k_{+6} = 14\ s^{-1}; \quad k_{-6} = 400\ s^{-1}; \quad K_7 = 2.7 \times 10^{-4}M.$$

These data imply that the stages determing the decrease of free energy during ATP hydrolysis by myosin are those connected with ATP binding (K_1 represents the binding and K_2, the conformational change of the M·ATP complex). The equilibrium constant of this process $K_1 \cdot K_2 \approx 10^{13}$ M^{-1} corresponds to a decrease in the standard free energy of the system by ∼18 kcal/mol. This value considerably exceeds the standard free energy of ATP hydrolysis under these conditions ($\Delta G^0 \approx -10$ kcal/mol [2.21]). Such an improbably large $K_1 K_2$ value is mainly due to the extremely low value of rate constant k_{-2} which is practically equal to the rate constant of ATP dissociation from the equilibrium stable ATP-myosin complex (the half time of this dissociation process according to these data must be ∼12 days). This is rather strange. The rate constant for the dissociation of an ATP analog (ATP-β, γ-NH, see above) was measured directly [2.19] and found to be 0.02 s^{-1}, i.e., higher than the estimated ATP value by 5 orders of magnitude.

All this forces us to prefer an alternative: postulates underlying the quasi equilibrium schemes of type (2) do not hold true in this case. The intermediates have not enough time to reach local equilibrium states, the paths of the direct and back reactions do not coincide.

Deshcherevesky understood quite clearly the limited value of the schemes of types (2.2,3) and the necessity to account for the finite duration of conformational changes [2.3,27]. He could not complete his research in this field.[1] (Footnote 1 see next page).

It is easy to understand that the schemes described above for actomyosin ATPase are based on the same postulates of classical physical chemistry as (2.2). This is also true for the analysis of the sliding filament model in the realm of the absolute reaction rate theory [2.28]. The data presented in this section allow us to express a doubt in the applicability of these postulates to the functioning of elementary chemico-mechanical energy transducers in the systems of biological motility.

2.2 Active Transport of Ions

Energy "taken money" of living matter, adenosine triphosphate (ATP) is spent in the course of many energy-accepting processes. The most important energy consumers under normal conditions (e.g., at not extremely intensive muscle activity) are the processes of active transfer of ions against their electrochemical potential gradients. The term "electrochemical potential" deserves in this case a special discussion (Chap.3). We shall assume, for the time being, that we are speaking about the maintenance of nonequilibrium concentration distribution of ions between phases separated by a membrane. Biological membranes are not ideal insulators, there is always a passive leakage of ions through specific ion channels or by means of specific carriers (this leakage sometimes rises significantly due to the local increase in membrane permeability, e.g., during the generation of a nerve pulse).

Continuous work of specific "pumps" is required to maintain this nonequilibrium distribution of ions. Such pumps are the Ca^{2+}-pump of sarcoplasmatic reticulum membranes, the Ca^{2+}, Mg^{2+}-pump of plasma membranes and erythrocytes, etc. The best studied and, probably, the most energy-consuming, the "sodium pump" (Na-pump) pumps Na^+ ions out of and K^+ ions into the cells. Na-pumps are localized in plasmatic membranes of all cells (this membrane represents a boundary between a cell and the surrounding medium) and determine such important processes as the generation and propagation of excitation potential, the regulation of transmembrane transfer of many biologically active compounds, the regulation of many enzymatic reactions.

In 1957 SKOU [2.29] found the functioning of the Na-pump to be connected with its ATPase activity: the pump is a membrane-bound lipoprotein, while the enzyme is Na-, K-dependent ATPase, (ATP hydrolyzing enzyme, adenosine triphosphatase), capable of catalyzing ATP hydrolysis in the presence of Na^+, K^+ and Mg^{2+} ions. The identity of Na-, K-ATPase isolated from the membrane and of the Na-pump which had been for many years the object of physiological and biochemical research, is now strictly established. As a result of investigations carried out in numerous laboratories of many countries in the last 20 years, the molecular structure of this enzyme and the basic biochemical peculiarities of its functioning have been clarified quite satis-

---

1 V.I. Deshcherevsky died in 1975 at the age of 36.

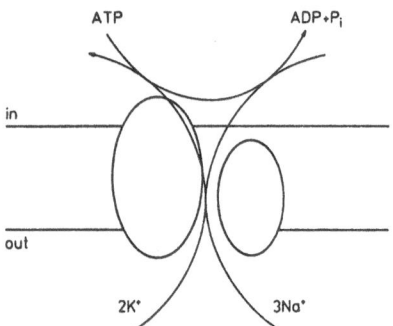

ATP   ADP+P$_i$

in

out

2K$^+$ 3Na$^+$

Fig.2.2. The functional unit of Na-, K-ATPase

factorily [2.30-33]. Although the native enzyme contains, as a minimum, four subunits, its functional unit is a dimer composed of one large subunit (a lipoporotein,
MW ≈ 100000 D) and one small subunit (glycoprotein, MW ≈ 50000 D). The large subunit
penetrates the membrane, and the small subunit is localized on the membrane inner
surface (Fig.2.2).

The stoichiometry of Na-, K-ATPase functioning corresponds to the transfer of
three Na$^+$ ions in exchange for two K$^+$ ions in response to the hydrolysis of one
ATP molecule. We have seen that in the case of muscle contraction the switching on
of an energy-accepting process (the formation of the myosin-actin complex and the
consequent development of the pushing force) accelerates the energy-donating process
of ATP-hydrolysis. The same phenomenon can be observed in the case of the sodium
pump: the interaction of the enzyme with transferred ions leads to ATPase activation.
The membrane-bound enzyme functioning requires that Na$^+$, Mg^{2+} and ATP should be at
the inner side, and K$^+$ at the outer side of the membrane. The ATPase activity of
isolated enzyme in solution is not equal to zero only in the presence of both the
Na$^+$ and the K$^+$ ions.

Many formal biochemical schemes of ATP hydrolysis by Na-, K-ATPase have been proposed. As a rule, these schemes can be subdivided into two types of mechanisms: with
sequential enzyme activation by transported ions (firstly by Na$^+$ and then by K$^+$
[2.34-37]) and with simultaneous concerted enzyme activation by these ions [2.31,38,
39]. A detailed analysis of these scheme can be found in [2.33] and in the original
publications cited above. These two extreme points of view have been recently com-
bined [2.40,41]. In all these schemes ATP hydrolysis leads to the formation of a
phosphorylated protein undergoing conformational transition, as a result of which
the exchange of Na$^+$ and K$^+$ ions takes place. As in the case of biochemical schemes
of ATP hydrolysis by myosin and actomyosin, the sequence of local chemical acts and
conformational changes is regarded as a traditional multistep chemical reaction with
totally reversible individual stages. The energy liberated during ATP hydrolysis is
assumed to be spent on providing for the enzyme conformational transitions.

It is, therefore, appropriate to describe here the attempts to reverse the sodium
pump, i.e., to use the Na-, K-ATPase as a catalyst of ATP synthesis from ADP and P$_i$.

The first work dealing with this problem was published in 1971 [2.42]. According to the authors, the energy source for ATP synthesis is in this case Na^+ and K^+ concentration gradients artificially created in the preparations of erythrocyte ghosts: immersion of erythrocyte ghosts containing K^+, P_i and ADP into the medium with high Na^+ concentration actually resulted in the exclusion of K^+ ions from erythrocyte stroma and in a not very significant ATP synthesis [incorporation of ^{32}P (P_i) into adenylates].

It was, however, shown later that Na-, K-ATPase is able to synthesize ATP in the absence of Na^+ and K^+ concentration gradients [2.43-46]. These interesting experiments were carried out in the following way. An isolated membrane enzyme preparation containing bounded K^+ ions was phosphorylated by P_i in the presence of Mg^{2+} ions. If the enzyme is then transferred into the medium with Na^+ ions and ADP, the K^+ ions are substituted with Na^+ ions and all the phosphate from phosphorylated enzyme is incorporated into ATP. The authors have assumed that the energy liberated by the substitution of K^+ ions with Na^+ ions is used for ATP synthesis. This substitution is accompanied by the conformational transition of the enzyme from its "K-form" into the "Na-form". The available experimental facts thus allow us to conclude that a typical process of the active transport of ions — the functioning of sodium pump — is actualized with the help of a molecular machine, a membrane-bound enzyme able to couple the energy-donating and energy-accepting chemical reactions.

2.3 Substrate Phosphorylation

Up to now only the bioenergetic processes in which the role of energy-donating act was played by ATP hydrolysis have been discussed. A corresponding free energy decrease is used to perform such energy-accepting processes as muscle contraction or the transport of ions against their electrochemical potentials. Let us now turn to the processes in which the ATP, ADP and P_i system undergoes energy-accepting transformations. In other words, let us consider the processes of ATP synthesis from ADP and P_i. In all intracellular reactions of ADP phosphorylation the energy-donating process is another chemical reaction, as a rule, that of oxidation. In textbooks on biochemistry the standard scheme of ATP formation is, therefore, usually written in the form

$$A_{ox} + B_{red} \longrightarrow ATP + H_2O$$
$$A_{red} + B_{ox} \longleftarrow ADP + P_i \quad . \tag{2.5}$$

As in all bioenergetic processes, the main problem in this case is the mechanism of energy coupling of chemical reactions.

Two types of ATP synthesis processes exist: substrate and membrane phosphorylation. The latter process takes place within organized membrane multi-enzyme structures of

mitochondria (oxidative phosphorylation) and chloroplasts (photosynthetic phosphorylation or simply "photophosphorylation"). Substrate phosphorylation, to be discussed in this section, is carried out by isolated, mainly glycolitic enzymes. It
can be easily realized in a solution in the absence of any supramolecular structures.

As an example of a substrate phosphorylation process let me discuss here the ATP
synthesis catalyzed by phosphoglycerokinase (designated henceforth as E_p - phosphorylating enzyme) caused by the energy-donating process of glyceraldehyde-3-phosphate
oxidation, with coenzyme NAD^+ catalyzed by glyceraldehyde-3-phosphate dehydrogenase
(this enzyme will be designated as E_0 - oxidizing enzyme). The overall reaction can
be written down as follows:

$$\text{glyceraldehyde-3-phosphate} \qquad + NAD^+ + H_2O \qquad E_0 \quad E_p \qquad ATP + H_2O$$

$$\text{3-phosphoglyceric acid} \qquad + NADH + H^+ \qquad ADP + P_i \tag{2.6}$$

The biochemical mechanism of this process was clarified in the brilliant research
carried out in Racker's laboratory in the 1950s. (A fascinating description of this
research can be found in [2.47]). E_0 forms a strong complex with oxidized coenzyme
NAD (probably through the S atom of one of the enzyme sulfohydril groups)

$$E_0 - S - H + NAD^+ \rightarrow E_0 - S - NAD + H^+ \quad . \tag{2.7}$$

The acyl-enzyme, in which the substrate is already oxidized, is formed as follows:

$$H - \overset{\overset{H}{|}}{\underset{\underset{CH_2OPO_3H_2}{|}}{C}} - OH + E_0 - S - NAD \longrightarrow E_0 - S \sim \overset{O}{\overset{\|}{C}} - \overset{\overset{H}{|}}{\underset{\underset{OH}{|}}{C}} - CH_2OPO_3H_2 + NADH \quad . \tag{2.8}$$

$$\text{acyl-enzyme}$$

Acyl-enzmye is already a high-energy compound: acylation is thermodynamically unfavorable.

The acyl-enzyme hydrolysis

$$E_0 - S \sim \overset{\overset{O}{\|}}{C} - \overset{\overset{H}{|}}{\underset{\underset{OH}{|}}{C}} - CH_2OPO_3H_2 + H_2O \longrightarrow$$

$$\longrightarrow E_0 - S - H + HO - \overset{\overset{O}{\|}}{C} - \overset{\overset{H}{|}}{\underset{\underset{OH}{|}}{C}} - CH_2OPO_3H_2 \qquad\qquad (2.9)$$

is accompanied by a decrease in the system free energy the excess of which is dissipated.

One of the end products of (2.6), 3-phosphoglyceric acid, is formed, and the enzyme is regenerated. Although reaction (2.9) is energetically favorable, it proceeds extremely slowly: the acyl-enzyme in a water medium is kinetically stabilized. The formation of 3-phosphoglyceric acid from glyceraldehyde-3-phosphate at catalytic quantities of enzyme does not, therefore, practically take place (the enzyme "turnover number" is very low).

At higher enzyme concentrations, when E_0 plays the role of a reagent and not of a catalyst, reaction (2.8) can proceed quantitatively. Although the formation of acyl-enzyme in a water medium is thermodynamically unfavorable, (2.8) still proceeds because simultaneously a thermodynamically favorable reaction of NAD^+ reduction is taking place. We shall return later to the meaning of the word "simultaneously" in this case.

The glyceraldehyde-3-phosphate oxidation, however, proceeds within the cell in the presence not only of NAD^+ but of P_i, as well. Inorganic phosphate increases the enzyme turnover number more that 150000 times. Reaction (2.8) is accompanied now not by acyl-enzyme hydrolysis with energy dissipation, but by process

$$E_0\text{-}S\sim\overset{\overset{O}{\|}}{C}\text{-}\overset{\overset{H}{|}}{\underset{\underset{OH}{|}}{C}}\text{-}CH_2OPO_3H_2 + H_3PO_4 \longrightarrow E\text{-}S\text{-}H + H_2O_3P\text{-}O\sim\overset{\overset{O}{\|}}{C}\text{-}\overset{\overset{H}{|}}{\underset{\underset{OH}{|}}{C}}\text{-}CH_2OPO_3H_2 \quad . \qquad (2.10)$$

The enzyme is regenerated and 1,3-diphosphoglycerate with a high-energy bond in position 1 is formed.

In this way a low-molecular high-energy compound is formed by utilizing the thermodynamically favorable process of NAD^+ reduction with glyceraldehyde-3-phosphate. The last, concluding stage of the process is the transfer of high-energy bonds onto the ADP catalyzed by E_p:

$$H_2O_3P - O \sim \overset{\overset{O}{\|}}{C} - \overset{\overset{H}{|}}{\underset{\underset{OH}{|}}{C}} - CH_2OPO_3H_2 + ADP \xrightarrow{E_p} HO - \overset{\overset{O}{\|}}{C} - \overset{\overset{H}{|}}{\underset{\underset{OH}{|}}{C}}CH_3OPO_3H_2 + ATP \quad . \qquad (2.11)$$

The algebraic sum of (2.7,8,10,11) corresponds, naturally, to (2.6). We shall not discuss the latter stage (2.11). It does not mean that the physical mechanisms of high-energy bond transfer between two molecules are absolutely clear. No new problems would be raised by such a discussion which could not be analyzed by considering the preceding stages.

In the presence of P_i the energy-donating reduction of E_o-bond NAD^+ thus ensures the energy-accepting process of glyceraldehyde-3-phosphate oxidation and phosphorylation accompanied by the formation of a high-energy bond. We can visualize this process as proceeding in one elementary act

$$
\begin{array}{c}
H \diagdown \quad \diagup O \\
C \\
| \\
H - C - OH \\
| \\
CH_2OPO_3H_2
\end{array}
+E_o\text{-S-NAD}+H_3PO_4 \longrightarrow E_o\text{-SH}+NADH+H_2O_3PO\text{-C-C-CH}_2OPO_3H_2
\qquad (2.12)
$$

and assume that under normal functioning conditions in the presence of P_i the acyl-enzyme does not exist as a kinetically independent unit. Reaction (2.12) is, in this case, an act of energy coupling of chemical reactions. Indeed, it is known that E_o is a "two-headed" enzyme with two main active centers, one of which is responsible for acylation (i.e., for the substrate binding and acyl-enzyme formation observed in the absence of P_i), and the other, for P_i binding and transfer.

We can certainly imagine the existence of an alternative mechanism, according to which 1,3-diphosphoglycerate formation proceeds in two consecutive elementary acts: acylation (2.8) and subsequent fast high-energy bond transfer (2.12). Whatever the concrete mechanism would be, it is clear that in both cases a thermodynamically unfavorable chemical process (glyceraldehyde-3-phosphate phosphorylation or E_o acylation) must take place owing to a simultaneously proceeding thermodynamically favorable process (NAD^+ reduction). In this case, the word "simultaneously" means that the time interval between the two elementary acts is so small that the energy liberated during the thermodynamically favorable process does not have enough time to dissipate into heat. Here, both processes are in fact taking place in the course of one elementary act (or, speaking in terms of the absolute reaction rate theory, they have one and the same activated complex). The problem consists in clarifying the physical mechanism that can explain the dissipationless energy transfer in a condensed phase in the course of a process involving three low-molecular compounds and at least two different groups in two active centers of a protein. We see that in this case a design of molecule dimensions enzyme E_o, also exists which is a molecular machine ensuring energy coupling of chemical reactions.

2.4 Membrane Phosphorylation

Membrane phosphorylation is probably the most important bioenergetic process. It is performed by specific intracellular organelles: mitochondria (oxidative phosphorylation) and chlorophlasts or chromatophores (photosynthetic phosphorylation). Oxidative phosphorylation is responsible for about 85% of the energy utilized within the cells of aerobic organisms, and during photophosphorylation a considerable part of light quanta energy absorbed by pigments of plants and photosynthetic bacteria is utilized.

Let us begin with oxidative phosphorylation in mitochondria. Food compounds after many metabolic chemical transformations in the majority of cell types transfer their reducing equivalents to two low-molecular compounds: nicotinamideadeninedinucleotide (NAD-H) and succinate. Chemical formulas of NAD-H and succinate and the schemes of their redox transformations are given in (2.13,14):

$$+ 2e^- + H^+ \tag{2.13}$$

$$HCOOC - CH_2 - CH_2 - COOH \rightleftharpoons HOOC - CH = CH - COOH + 2e^- + 2H^+ \quad . \tag{2.14}$$

succinic acid   fumaric acid

Mitochondria are intracellular organelles whose membranes contain "electron carriers" and enzymes catalyzing ATP synthesis from ADP and P_i (ATP synthetase which contains two macromolecular complexes F_0 and F_1, it is sometimes called "coupling factor"). Electron carriers form the so-called electron transport chains (ETC). Figure 2.3 depicts a mitochondrion. An elongated ellipsoidal particle (long axis ~3 μm, short axis ~1 μm) is surrounded by two membranes: a smooth outer membrane, and an inner membrane with multiple cristae forming a rather complicated pattern. The space within the inner membrane is filled by a protein-containing "matrix". The inner membrane of one mitochondrion contains about 1500 ETC.

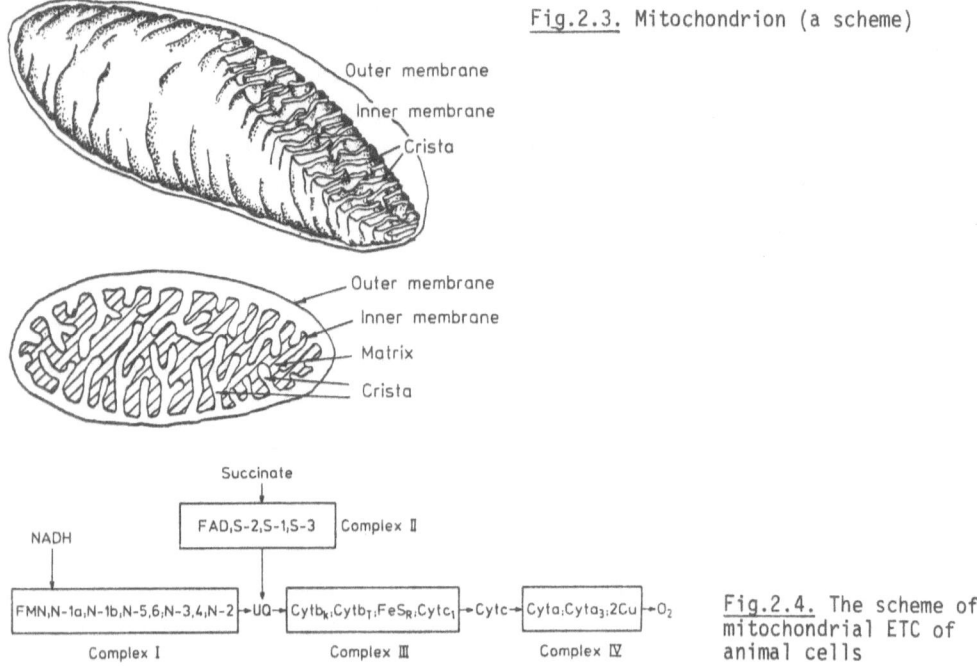

Fig.2.3. Mitochondrion (a scheme)

Outer membrane

Inner membrane

Crista

Outer membrane

Inner membrane

Matrix

Crista

Succinate

| FAD,S-2,S-1,S-3 | Complex II |

NADH

| FMN,N-1a,N-1b,N-5,6,N-3,4,N-2 | →UQ→ | $Cytb_K$,$Cytb_T$,FeS_R,$Cytc_1$ | →Cytc→ | $Cyta$,$Cyta_3$,2Cu | →O_2 |

Complex I   Complex III   Complex IV

Fig.2.4. The scheme of mitochondrial ETC of animal cells

Figure 2.4 depicts animal mitochondrial ETC; ETC can be subdivided into several complexes usually called "Green complexes". Fragmentation of mitochondrial membranes and isolation of the functioning oligoenzyme complexes have been carried out in the laboratory of GREEN [2.48]. Subdivision of carriers into complexes is very convenient, but there is no conclusive evidence of the existence of corresponding morphologically determined formations within ETC. Two ETC substrates, NADH and succinate, interact with ETC through the first and second Green complexes, respectively (these complexes are often called NADH-dehydrogenase and succindehydrogenase). These complexes contain flavoproteins (FP) whose active centers are FMN, flavinemononucleotide (complex 1), and FAD, flavineadenine-dinucleotide (complex 2), as well as nonheme-iron proteins (Fe-S-Proteins), denoted "N" (complex 1) and "S" (complex 2). The structures and schemes of redox transformations of these carriers are given in Fig.2.5. Two paths of electron transport from NADH and succinate join in the "ubiquinone pool" (UQ). Ubiquinone, a low-molecular carrier (also called "coenzyme Q_{10}; or simply "coenzyme Q"), can be easily extracted from mitochondria by organic solvents. It transfers reducing equivalents farther along the ETC. Figure 2.6 schematizes redox and acid-base transformations of the most simple quinone. The reduced protonated and oxidized forms of ubiquinone are hydrophobic and lipid-soluble, and in semireduced ion-radical form the quinone "head" of an ubiquinone molecule becomes hydrophilic. These properties probably play an essential role in electron transport: ubiquinone transfers electrons between centers localized on different sides of the inner membrane of mitochondria, simultaneously ensuring transfer of protons.

18

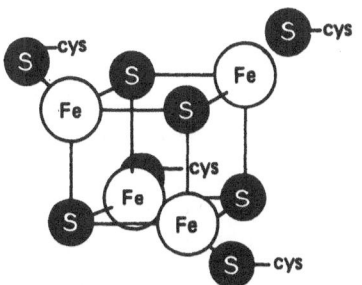

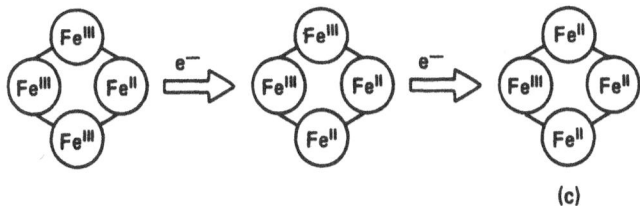

Fig.2.5. Structures and redox transformation schemes of flavine (a), binuclear Fe-S-center (b) and tetranuclear Fe-S-center (c)

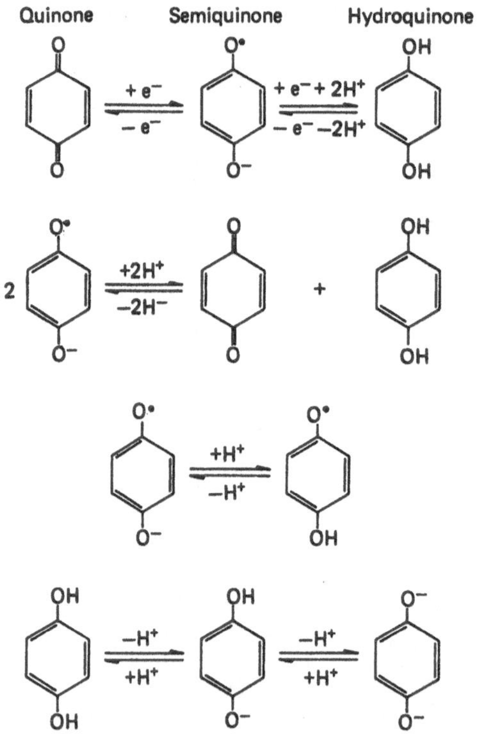

Quinone   Semiquinone   Hydroquinone

Fig.2.6. The schemes of redox and acid-base transformations of benzoquinone

The third Green complex contains one-electron carriers: cytochromes b_{562}, b_{566}, c_1 and nonheme-iron protein with an iron-sulfur active center Fe-S_R.[2] The structures and schema of redox transformations of cytochrome active centers can be seen in Fig.2.7. The chemical structures of all b-type cytochromes are identical. This is also true for cytochromes of the c and a types. In the ETC scheme (Fig.2.7) cytochrome c is situated between complex 3 and 4. This carrier can be easily extracted by hypertonic salt solutions, and, therefore, can remain outside the complexes during their isolation. This cannot, however, be regarded as evidence of it being a mobile carrier not fixed within a membrane framework. It is well known that native conformational state of cytochrome c is stabilized after its inclusion into ETC membrane [2.52]. Complex 4 contains cytochromes a and a_3, and copper ions forming a complex enzyme, cytochrome-c-oxidase, the terminal ETC component of all aerobic organisms.

Excellent reviews concerning the structure and properties of this enzyme can be found in [2.52,53]. Cytochrome-c-oxidase is a complex subunit protein containing 20% of the phospholipids obligatory for its functioning. The most probable number of subunits is 7. The molecular complex formed by these subunits and phospholipids

---

2 Subscripts "562", "566" refer to the wavelengths of absorbtion bands in reduced states (in nm). According to Chance's classification cytochromes b are designated as b_k and b_t [2.49]. Fe-S_R center is usually called "Rieske center" [2.50].

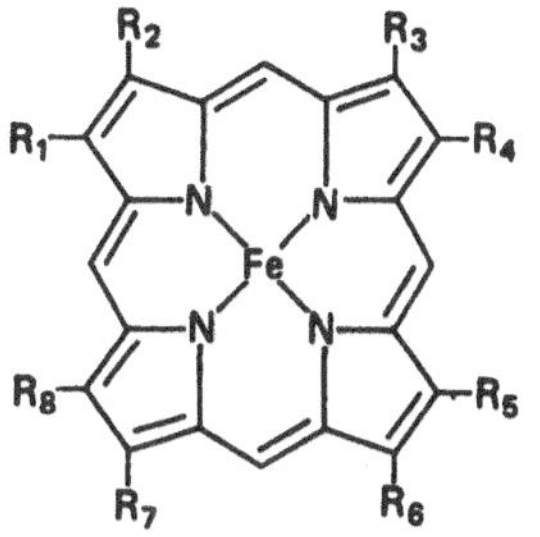

Fig.2.7. Redox transformations of cytochrome active centers

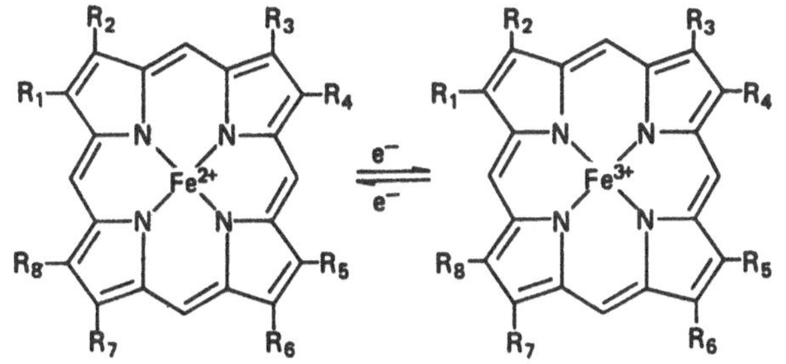

contains two copper ions and two hemes (a and a_3) which seem to be chemically identical after isolation but differ greatly in their properties in native preparations.

Two peculiarities of mitochondrial ETC organization must be emphasized. The first
one is the very large concentration of metal atoms, mainly of iron. One ETC contains
about 50 iron and 2 copper atoms. The low-temperature EPR studies of organic free-
radical centers in mitochondria (flavo- and ubisemiquinones) have shown that every
flavine and quinone group in ETC is localized in the neighborhood of one of the ETC
iron atoms (cytochromes or nonheme-iron proteins) [2.54]. This makes the solution of
the problem concerning the physical mechanism of electron transfer between ETC components rather trivial: the overlap of electron orbitals neighboring centers (either
directly or via intermediate bridges) can easily ensure observed electron transport
rates if the electron levels of centers are properly adjusted (see also Chap.7 in
[2.55]).

The second peculiarity is the rigid fixing of carriers within the membrane. In
each state of ETC not only the position, but even the orientation of the active
centers of most carriers are strictly fixed: the angles between the membrane plane
and the planes of cytochrome a and a_3 hemes in cytochrome-c-oxidase [2.53,56], the
angles between the membrane planes and the axes of bi- and tetranuclear iron-sulfur
centers in complexes 1 and 2 [2.57]. Even some low-molecular carriers manifest
pronounced orientational anisotropy. The role of electron acceptor to succindehydrogenase complex 2 is played in some systems by a pair of exchange-coupled ubiquinone
molecules [2.58]. It was found that the vector connecting the aromatic ring centers
of this pair (the "centers of gravity" of the density distribution of unpaired electrons in the semiquinone half-reduced state of the carrier) cannot deviate from normal to the membrane plane by more than 15° [2.57].

21

The overall reaction of the NADH oxidation in mitochondrial ETC can be written as follows:

$$NADH + H^+ + 1/2O_2 \rightarrow NAD^+ + H_2O \quad . \tag{2.15}$$

The corresponding free-energy change in the conditions of a cell ($[NADH]/[NAD^+] \approx 1$, $P_{O}2 \approx 0.2$ atm, pH ≈ 7.0) is [2.59]

$$\Delta G \approx -51 \text{ kcal/mol} \quad . \tag{2.16}$$

The overall reaction of succinate oxidation in mitochondrial ETC

$$HOOC - CH_2 - CH_2 - COOH + 1/2\ O_2 \rightarrow HOOC - CH = CH - COOH + H_2O \tag{2.17}$$

in the conditions of a cell corresponds to the free-energy change

$$\Delta G \approx -36 \text{ kcal/mol} \quad . \tag{2.18}$$

The passage of two electrons through ETC from NADH to O_2 from succinate to O_2 in the undamaged mitochondria leads to the synthesis of 3 or 2 ATP molecules, respectively. The system is thus able to spend for the formation of one ATP molecule $\sim 51/3 = 17$ kcal during NADH oxidation and $\sim 36/2 = 18$ kcal during succinate oxidation.

The free energy required for the formation of one ATP molecule can easily be estimated with the help of a well-known formula of equilibrium thermodynamics:

$$\Delta G = \Delta G^o + 1.36\ \frac{[ATP]}{[ADP][P_i]} \quad , \tag{2.19}$$

where ΔG^o is the standard free energy of ADP phosphorylation (water concentration is assumed to be constant and accounted for in the ΔG^o term). In state 4 of mitochondrial ETC (see below) $\Delta G \approx 14.5$ kcal/mol [2.60]. We thus see that for both ETC substrates their oxidation supplies enough energy to ensure ATP synthesis.

A more minute analysis of electron transfer thermodynamics in mitochondrial ETC leads, however, to another conclusion. It is generally accepted today that ETC contains local "coupling sites", i.e., "carriers-transformers" and that electron transfer through these sites is accompanied by a decrease in free energy which is used for ATP synthesis. These coupling sites can function independently and are localized in the 1st, 3rd and 4th Green complexes. Carrier-transformer in the first complex is the N-2 center [2.61], in the third, cytochrome b [2.62], and in the fourth, the pair of cytochromes a and a_3 [2.53, pp.63-65].

According to equilibrium redox potential values of the ETC carriers, the free energy changes accompanying the passage of one electron through coupling sites in the first, third and fourth Green complexes are ~ 240, ~ 280 and ~ 120 mV [2.66-68] respectively. The passage of two electrons through three coupling sites can thus provide not much more than ~ 1300 mV or ~ 30 kcal/mol for ATP synthesis. It is evidently not enough: in the cell conditions of a cell no less than 43 kcal/mol are necessary [2.60]. This "energy crisis" can be overcome in only tree ways:

1) refusing to use the formulas of the (2.19) type in bioenergetics; 2) assuming that measured values of equilibrium redox potentials bear no direct relation to the intracellular energy transduction processes; 3) assuming that no local independent coupling sites exist. We shall see later (Sect.5.4.2) that the former one is the right way.

An appropriate classification of mitochondrial ETC states was suggested in CHANCE's laboratory [2.69]. According to this classification, isolated mitochondria can exist in five different ETC states. State 1 is realized in the absence of added respiration and phosphorylation substrates. ETC carriers (cytochromes, flavoproteins, nonheme-iron proteins) are partly reduced.

In *coupled* mitochondria (i.e., mitochondria in which the respiration leads to ATP synthesis) not only the energy-accepting process depends on the energy-donating one, but vice versa: switching on the energy-accepting process influences essentially the energy-donating process. If phosphorylation substrate (ADP) is added to mitochondria in state 1, the rate of electron transfer from NAD-H or succinate to oxygen (the tissue respiration intensity) increases and is now determined by respiration substrates (state 2). As in the case of muscle contraction, switching on the load leads to the acceleration of the energy-donating process. All ETC carriers are oxidized. If we now add the respiration substrates, mitochondria pass into the active state 3, the carriers are again partly reduced, and electron transport rate is determined by the enzymatic activity of the phosphorylation system and by the velocity of substrates' penetration through the outer mitochondrial membrane. When a greater part of ADP transforms into ATP, or if a large excess of ATP is added the state of respiratory control arises (state 4), respiration rate decreases and, according to widely accepted views, it is now determined by the [ADP]/[ATP] ratio. By changing this ratio one can switch over mitochondria state 4 to state 3 and vice versa. In the absence of O_2 mitochondria are in a state of anaerobiosis (state 5) in which all carriers are reduced.

In coupled mitochondria the ADP phosphorylation is ensured by the energy liberated within the ETC in the course of electron transfer from NADH or succinate to O_2:

$$NAD\text{-}H + 1/2O_2 + H^+ \longrightarrow 3ATP + 3H_2O$$
$$NAD^+ + H_2O \longleftarrow 3ADP + 3P_i \tag{2.20}$$

$$(CH_2COOH)_2 + 1/2O_2 \longrightarrow 2ATP + 2H_2O$$
$$(CHCOOH)_2 + H_2O \longleftarrow 2ADP + 2P_i \quad . \tag{2.21}$$

Energy-donating and energy-accepting reactions can be uncoupled: in uncoupled mitochondria the energy liberated due to NADH or succiate oxidation dissipates, and ATP is not synthesized. Mitochondria pass into the uncoupled state as a result of even weak damage: heating up to $50^\circ C$, repeated freezing and thawing procedure, aging.

The coupling ability, i.e., the ability to synthesize ATP owing to electron transfer processes in ETC, is lost much more easily than the respiration activity, i.e., the ability to oxidize NADH or succinate by molecular oxygen. There also exists a large group of "uncouplers" whose addition leads to the uncoupling of mitochondria.

In an uncoupled state, when energy-donating processes of electron transfer are "running idle", their rate greatly increases. It means that uncoupled mitochondria are characterized not only by the absence of ATP synthesis. The ATP formation can be abolished in coupled mitochondria too, in the absence, e.g., of phosphorylation substrates (ADP or P_i). In the latter case the electron transport rate decreases, but shutting off the ATP synthesis due to uncoupling leads to an increase in the electron transport rate.

Detailed kinetic studies carried out in CHANCE's laboratory [2.70,71] have shown that uncoupling accelerates electron transfer through coupling sites. It is these stages that determine the rate of electron transport process in coupled mitochondria.

Let us consider now the processes of electron transport and phosphorylation in photosynthetic systems. For our purpose, i.e., for the following analysis of energy transduction physics, it is quite sufficient to limit our discussion to the main characteristics of these processes in the chloroplasts of green plants. A description of energy-transducing membranes in chromatophores of photosynthetic bacteria and of their functioning can be found in many excellent reviews [2.72,73].

The overall chemical equation of photosynthesis in green plants can be written as follows:

$$nCO_2 + nH_2O \rightarrow (CH_2O)_n + nO_2 \quad . \tag{2.22}$$

This overall process includes many steps, and can be rather arbitrarily subdivided into the light and the dark stages. In the course of the light stage ATP is formed, $NADP^+$ reduced, and water oxidized with the liberation of molecular oxygen:

$$H_2O + NADP^+ + 2ADP + 2P_i \rightarrow 1/2O_2 + NADPH + 2ATP + H^+ \quad . \tag{2.23}$$

The products of the light stage, NADPH and ATP, are used in the dark stage for CO_2 reduction and the synthesis of carbohydrates in the course of the enzymatic reaction of the Calvin cycle [2.74]. Process (2.23) cannot, strictly speaking, be defined as a "light stage". Only the primary acts of light absorption by the pigments of the "light-harvesting matrix" (see below), excitation migration to reaction centers, and charge separation with the formation of primary oxidizer and reducer, are true "light stages". The subsequent steps of electron transfer and coupled phosphorylation are the very same dark processes as the corresponding processes within the mitochondrial membrane. We shall, however, follow the established tradition and designate the overall reaction (2.23) as a "light stage".

The smallest structural and functional unit able to perform the light stage (2.23) in chloroplasts is, probably, thylakoid—a discoidal closed vesicle with

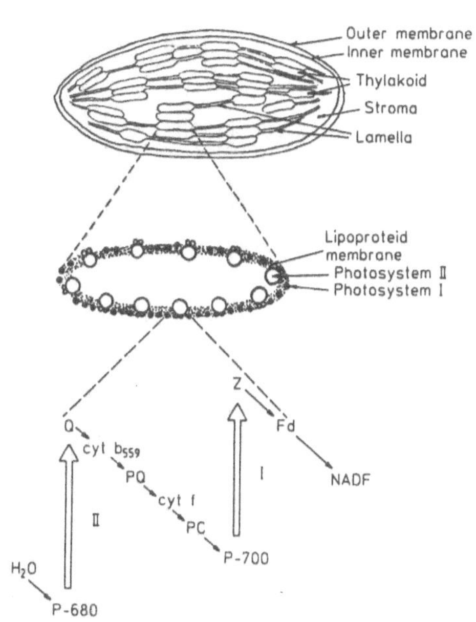

Outer membrane
Inner membrane
Thylakoid
Stroma
Lamella

Fig.2.8. Chloroplast, thylakoid, and the ETC scheme of higher plants

Lipoproteid membrane
Photosystem II
Photosystem I

Z
Q
Fd
cyt b$_{559}$
PQ
I
NADF
cyt f
II
PC
P-700
H$_2$O
P-680

diameter ~500 nm. One chloroplast contains about 10^3 thylakoids. The scheme of a thylakoid is shown in Fig.2.8. The membrane of one thylakoid contains ~10^5 molecules of light-harvesting pigment (chlorophyll) and ~200 ETC. Every ETC includes two photosystems, and the transfer of one electron through ETC requires the absorption of two light quanta. The major fraction of chlorophyll (and of certain other pigments) forms the so-called light-harvesting matrix: the singlet excitation formed as a result of light quantum absorption by any matrix molecule of a given photosystem migrates to the corresponding reaction center where a photochemical act of charge separation takes place. The charged centers — electron donors and electron acceptors — are thus formed and subsequently discharge in the course of electron transfer between ETC components. Reaction centers of photosystem 1 (PS1) are, most probably, chlorophyll a dimers, also called "P700" centers (according to the position of their absorption band: $\lambda_{max} \approx 700$ nm). A similar center in photosystem 2 (PS2) is called P680. One reaction center serves 200-300 chlorophyll molecules of light-harvesting matrix. The sequence of fast stages: light absorption, excitation migration and charge separation up to the formation of more or less stable trapped electrons and holes, which play the roles of electron donors and acceptors for the initial steps of electron transfer through ETC, has been studied in detail by various physical methods [2.76,77].

Figure 2.8 shows the scheme of electron transfer in ETC of higher plants. The primary electron donors of PS1 reaction center (P700), cytochrome f and plastocyanine, receive electrons from P680, the active center of water-splitting PS2, through a set of carriers. In the course of noncycling electron transfer (and under normal conditions noncycling electron transport represents the main path of photosynthesis in green plants), the transfer of one electron from the initial donor

(H_2O) to the final acceptor ($NADP^+$) thus requires the absorption of two light quanta with the overall energy of ~3.65 eV, only ~1.12 eV of which is spent on electron transfer, the remainder ensuring ATP synthesis or dissipating. The rates of individual electron transfer stages are well known mainly due to the research carried out in the laboratory of WITT [2.77,78].

It is generally accepted now that in noncyclic ETC there are two coupling sites, and that the ATP yield in a steady state corresponds to the synthesis of two ATP molecules per pair of electrons transferred from H_2O to $NADP^+$. The light stage of photosynthesis can thus be represented in the form of two coupled chemical reactions:

$$H_2O + NADP^+ \quad\quad\quad\quad 2ATP + 2H_2O$$

$$O_2/2 + NADPH + H^+ \quad\quad\quad 2ADP + 2P_i \quad . \quad\quad\quad\quad (2.24)$$

The relationship between electron transport and phosphorylation processes in the case of photophosphorylation is the same as in the case of oxidative phosphorylation. The state of respiratory control in mitochondria corresponds to the state of photosynthetic control in chloroplasts.

It is now universally recognized that redox transformations of ETC carriers in mitochondria, chloroplasts, and chromatophores do not lead directly to ATP synthesis but to the formation of an "energized" intermediate, called "primary macroerg". This was postulated by all theoretical concepts of membrane phosphorylation differing from one another only in the assumptions concerning the nature of this intermediate "energized" state: chemical compound, transmembrane electrical potential, the difference between concentrations of certain ions (e.g., protons) on both sides of a membrane, the conformationally changed state of a membrane, etc. The formation of a primary macroerg not identical to ATP was, however, first shown in the early 1960s by two groups of scientists independently [2.79,80], and then verified and studied in detail in several laboratories [2.81]. Illumination of chloroplasts in the absence of ADP and P_i leads to the formation of a certain energized compound or state $\widetilde{X}$ which is capable of ensuring the ATP synthesis in the dark after ADP and P_i addition. The $\widetilde{X}$ formation is accompanied by changes in light-scattering properties of chloroplast suspension, indicating membrane conformational changes.

Kinetic characteristics of the rise and decay of these changes differ, however, from those of $\widetilde{X}$ rise and decay. These membrane conformational changes are, probably, secondary processes induced by $\widetilde{X}$ formation. $\widetilde{X}$ cannot be formed at temperature lower than $-13^\circ C$ although photoinduced electron transport proceeds down to $-30^\circ C$. In the absence of ADP and P_i the half-life time of $\widetilde{X}$ in the dark is 0.5-1 s at $20^\circ C$ and 45 s at $0^\circ C$. When $\widetilde{X}$ is formed during illumination it can synthesize in the dark up to ~30 ATP molecules per one ETC. It means that during illumination every coupling site performs its work several times.

Most specialists agree now that the mechanisms of membrane phosphorylation in ETC of mitochondria and of chloroplasts are identical. The formal general scheme of membrane phosphorylation can, therefore, be written as follows:

$$A_{red} + B_{ox} \quad\nearrow\!\!\!\nwarrow \quad X^{\sim}\!\!-\!\!\nearrow\!\!\!\nwarrow \quad ATP + H_2O$$
$$A_{ox} + B_{red} \quad\swarrow\!\!\!\searrow \quad X\!\!\swarrow\!\!\!\searrow \quad ADP + P_i \qquad . \qquad\qquad (2.25)$$

In all membrane phosphorylation systems ATP synthesis is ensured by a specific enzyme, a "coupling factor", designated as F_1 in mitochondria, and as CF_1 in chloroplasts. These complex proteins are very similar. They are particles in 90-100 Å in diameter in with MW of 325000-350000 D containing five subunits. F_1 are localized at the inner side of the mitochondrial inner membrane and CF_1 at the outer side of the thylakoidal membrane in chloroplasts. A detailed description of F_1 and CF_1 structures and properties can be found in reviews [2.82].

F_1 and CF_1 factors are often called ATP-synthetase. In coupled mitochondria and chloroplasts this enzyme catalyses the ATP synthesis from ADP and P_i utilizing the energy liberated as a result of redox transformation reactions at coupling sites. In order to make, e.g., CF_1 catalyse the back reaction, i.e., function as an ATPase, specific activation is required (heating, trypsin treatment, acid-base transition, etc.). This activation is probably reduced to hydrolytic dissociation or conformational change of a certain "inhibitory" polypeptide that in normal conditions is part of a protein globule [2.83].

A series of extremely interesting papers by a group of Spanish scientists clarifying the functional mechanism of mitochondrial coupling factor has been recently published [2.84-87]. The authors have worked with isolated and purified factor F_1 (F_1 isolation leads to the appearance of its ATPase activity). They have found that catalytic activity of this mitochondrial F_1 ATPase changes reversibly under the action of redox agents. After reduction with hydrosulfite the ATPase activity increases but the characteristic ATPase property to be activated by bicarbonate is lost almost completely. The sample reoxidation by 2,6-dichlorophenolindophenol restores the initial state: ATPase activity drops to its initial level, and the ability to be activated by bicarbonate is restored. Some of the these results have been confirmed in our laboratory with the CF_1 factor isolated from chloroplast [2.90].

Appearance of free-radical centers in the course of oxidative phosphorylation in mitochondria was described in the work of KAJUSHIN and his co-workers [2.88-89]. One of the most frequently applied methods in membrane phosphorylation research is the investigation of the action of various uncouplers [2.91].

2.5 Formulation of the Main Physical Problems in Bioenergetics

In the preceding sections a phenomenological description of the most important bioenergetic processes has been given. The analysis has shown that in all cases we are
dealing with one and the same phenomenon: energy coupling between two chemical reactions—the energy-donating and the energy-accepting ones. The first main problem
of bioenergetics is, therefore, the mechanism of energy coupling of chemical reactions. Analysis of solutions, possible in principle, of this problem is given in the
next chapter. We shall see that this analysis will induce us to touch upon the
question of the theoretical justification of using in this case the classical physico-chemical approaches (those of chemical thermodynamics and chemical kinetics)
developed many years ago to describe and interpret chemical reactions in gaseous
phase and dilute solutions.

I shall now try to describe the situation presently existing among scientists
working in the field of bioenergetics. It can be done using the materials presented
at the extremely interesting conference "Mechanisms of Energy Transduction in Biological Systems" and organized by New York Academy of Sciences in 1973 [2.92].

In his letter to the participants, P. Mitchell who was later awarded the Nobel
Prize asked them to pay special attention in the discussion to the already existing
biochemical approaches and not to allow the discussion time to be occupied with medi-
tations concerning physical theories about direct interaction between "supermole-
cules". There is a widespread tendency to contrast "positive biochemical facts" with
"speculative physical theories". It was already stressed in the Introduction that
in modern science one can very rarely encounter pure facts unspoiled by a theory.
Even such simple statements as "the measured equilibrium constant of this reaction
is equal to...", or "the measured activation energy is equal to..." in reality reflect not experimental facts but the acceptance of certain postulates underlying
certain theoretical generalizations which are true only under strictly defined conditions (in these cases the mass action law and the Arrhenius approach to the elementary act of chemical transformation).

During the above-mentioned conference the position of "sceptical physicists" was
presented by the late McCLARE who in his report [2.93] and in his comments during
the general discussion [2.94] clearly showed that the principal questions, the
answers to which are required if we want to understand the mechanisms of bioenerge-
tical processes, are physical questions, and cannot be answered by means of any biochemical schemes. According to McClare, the main physical problem can be formulated
as follows: how can the energy liberated during an energy-donating process be stored
long enough to be used in an energy-accepting process if these processes take place
in a condensed phase? His postulates concerning the formation of long-living vibrational excitations have received rather convincing criticism. This, however, is not
very important. The situation here is almost the same (not comparing, of course, the
scopes of the problems) as it was formulated by the prominent Soviet physicist

Mandelstam apropos of the famous discussion between Bohr and Einstein in the thirties concerning the foundations of quantum mechanics. During this discussion Einstein suggested various mental experiments in order to bypass Heisenberg's principle of uncertainty, and Bohr in his answers explained why Einstein's reasoning was erroneous. At weekly seminars in the Physical Institute of the USSR Academy of Sciences, Mandelstam analyzed the schemes of experiments suggested by Einstein and, anticipating Bohr, convincingly showed their concrete errors. His co-workers wondered why he did not publish the results of this analysis. Mandelstam answered: "What for? If I can see it, so can Einstein. The concrete errors are in this case of no importance. What is important is the fact that Einstein is feeling that something is wrong here." (As recounted to me by the late Dr. S.M. Raysky.)

Description of the most important bioenergetic processes has shown us that in all cases energy transduction (i.e., the energy coupling of chemical reactions) is actualized with the help of specific macromolecular constructions, by machines of molecular dimensions. The second main problem of bioenergetics is, therefore, the development of a theory explaining the functioning of molecular machines. Before analyzing these physical problems it is appropriate to discuss the existing theoretical approaches to the mechanism of the best studied and, probably, the most important bioenergetic process—membrane phosphorylation.

3. Membrane Phosphorylation:
Chemiosmotic Concept and Other Hypotheses

In this chapter we shall first consider and, as far as possible, formulate precisely
the theoretical approaches to the mechanism of membrane phosphorylation proposed
during the last 30 years. Special attention will be given to Mitchell's chemiosmotic
concept, the citation rating of which is, probably, the highest in the history of
natural sciences. We shall then analyze the physical meaning of certain notions
used in various concepts of membrane phosphorylation. In the last part of this chapter the experimental data pro and contra the chemiosmotic concept will be summed
up and analyzed. I shall not deal here with the relaxation concept since it will be
discussed in Chap.5.

3.1 Survey of Existing Hypotheses

There are numerous reviews in current literature where various concepts of membrane
phosphorylation are compared and analyzed [3.1-7]. As already mentioned in Sect.
2.4, it can be considered as established that redox transformations of carriers at
coupling sites result in a certain "primary macroerg" $X^\sim$ being formed, and that
"discharge" of $X^\sim$ ensures the energy-accepting reaction of ATP synthesis (2.25). As
a matter of fact, all the hypotheses concerning the mechanism of membrane phosphorylation differ in the understanding of the nature of $X^\sim$.

Almost every one of the existing hypotheses (and there are many of them) can be
referred to one of the following three groups: chemical, conformational, and
chemiosmotic concepts. The first to be suggested was the chemical concept. In its
present form this concept was formulated by SLATER [3.8]. This concept is based on
the assumption that energy, as in the case of substrate phosphorylation, is stored
in the form of a covalent high-energy bond between one of the ETC electron carriers
(designated below as a "carrier-transformer" and denoted as T) and a ligand of un-
known nature (denoted below as I).

In the form suggested by KLINGENBERG [3.9], Slater's scheme can be written as
follows:

$$T_{red} + I \rightleftharpoons T_{red} - I \tag{3.1}$$

$$T_{red} - I + B_{ox} \rightleftharpoons T_{ox} \sim I + B_{red} \qquad (3.2)$$

$$T_{ox} \sim I + P_i \rightleftharpoons I \sim P_i + T_{ox} \qquad (3.3)$$

$$I \sim P_i + ADP \rightleftharpoons I + ATP \qquad . \qquad (3.4)$$

Carrier reduced states are often denoted as TH_2 and BH_2 [2.8,9]. Participation of hydrogen atoms in redox transformation is, as a matter of fact, not obligatory. Therefore the reduced or oxidized state of a carrier will be designated by means of corresponding indices. In this scheme it is assumed that energy is liberated in the course of carrier-transformer oxidation. If the energy-donating act turns out to be not the oxidation but the reduction of T, an appropriate modification of the scheme presents no difficulty.

Only in its reduced state T is assumed to be capable of forming a complex with ligand I [reaction (3.2)]. The liganded form of T is constructed in such a way that its oxidation by the neighboring ETC carrier B_{ox} is not accompanied by an energy decrease but transforms the T-I bond into a high-energy one. Reactions (3.3) and (3.4) lead to high-energy bond transfer and ATP formation, at least one of these reactions being enzymatic and requiring ATP-synthetase.

The chemical concept provides a natural explanation of the existence of respiratory control: electron transfer from T to the next carrier cannot take place without the simultaneous formation of "primary macroerg" $T_{ox} \sim I$, which, in its turn, would be blocked without reactions (3.3,4) that shift the equilibrium (3.2) to the right. A new version of the chemical concept has been proposed recently by LICHTENSTEIN and SHILOV [3.10].

For many years the chemical concept was questioned on account of two arguments. The first argument states that despite painstacking search nobody could find any intermediates [TI and IP in scheme (3.1-4)], such as those found, e.g., in the study of Na-pump (Sect.2.2). One can raise an objection against this argument: concentrations of intermediates can be negligibly low. This objection is, however, not very convincing. The chemical concept is based on equilibrium thermodynamics: macroerg formation is ensured by the displacement of reversible reaction (3.2) to the right due to the removal of the intermediates in the course of reactions (3,3,4). In a steady state near to equilibrium (e.g., in a fourth functional state of mitochondria with increased [ATP]/[ADP] ratio, see Sect.2.4) the concentrations of intermediates cannot be negligibly low.

The second argument implies that according to the chemical concept formulated above, ATP synthesis results from one act of carrier-transformer redox transformation. At the same time, in the ETC of chloroplasts and mitochondria there is no pair of neighboring carriers, the difference between whose redox potentials is high enough to ensure ATP synthesis in cell conditions (Sect.2.4). Schemes were therefore proposed in which either the transformer is a complex of two one-electron carriers, and the energies of their redox transformations add up to form a high-energy bond [3.1],

31

or the same carrier-transformer must undergo two consecutive redox reactions to synthesize ATP, i.e., the energies liberated during consecutive passages of two electrons must somehow add up [3.11].

I shall not analyze these schemes in greater detail. They do not answer the main question of the physical mechanism of coupling. It is easy to see that in the realm of the chemical concept the only difference between membrane and substrate phosphorylation lies in the fact that the redox reaction in solution is replaced by the act of electron transfer between neighboring ETC carriers fixed within a membrane. It is tacitly assumed that once membrane phosphorylation has been reduced to substrate phosphorylation everything becomes absolutely clear.

In 1962 TEMKIN and I examined a chemical scheme of membrane phosphorylation in which a possible principle of the coupling between the redox reaction in ETC and the acid-base reaction of ATP synthesis was proposed [3.12]. This scheme was based on the assumption that the adenine group in ADP takes part in the undergoing electron transfer in coupled ETC redox transformations. One-electron reduction of adenine ring

results in the appearance of the radical-anion form of adenine residue and is accompanied by an increase in the basicity of its amino group, which can reach the basicity value characteristic of aliphatic amines.

It easily binds a proton ($-NH_2 + H^+ \rightarrow -NH_3^+$) and then reacts with any of the electrophilic group, e.g., P_i. After that the formation of phosphate-ester bond between ADP β-phosphate group and P_i attached to the $-NH_3^+$ group of the adenine ring proceeds via a monomolecular reaction with almost no energy expenditure (because of the additional protonic charge). The oxidation of this radical-anion, i.e., the electron transfer farther into ETC, makes the adenine amino group aromatic again, its bond with P_i breaks down, and the ATP phosphate-ester bond becomes a high-energy one.

It was, however, doubtful that phosphate participated directly in the formation of the primary chemical high-energy compound [3.13]. SKULACHEV has suggested a carboxyl scheme, in which an increase in amino group basicity led to the formation of an amide bond with one of the carboxyl groups of the coupling enzyme [3.14,15]. The "discharge" of this primary macroerg should in Skulachev's scheme ensure ATP synthesis. Various possible schemes of ATP formation in the realm of chemical concept are minutely described in great detail in SKULACHEV's comprehensive monograph [3.16].

Let us now turn to Mitchell's chemiosmotic concept. The popularity of this hypothesis is startling. Almost every scientific paper concerning bioenergetic problems is concluded either with "our results thus confirm Mitchell's concept" or "our results thus contradict Mitchell's concept." In many papers the chemiosmotic concept is regarded as having been proved once and for all, while in others it is regarded as having been disproved once and for all. The first version of the chemiosmotic concept was published in 1961 [3.17]. Since then the details of the proposed schemes have been repeatedly changed in the papers of MITCHELL [3.18-20], not to mention the versions in the papers of his followers.

It is impossible and senseless to expound all the concrete biochemical details of these schemes and the history of their changes. Minute account of the chemiosmotic concept can be found in excellent reviews published by several authors [3.3,4,21, 22]. I shall, therefore, confine myself here to the formulation of the main propositions underlying this concept, which will help us to analyze later its physical validity. The essence of Mitchell's approach comes to three basic postulates.

1) The membrane phosphorylation process can be realized only in a membrane that forms a closed vsicle. The membrane separates the inner space of this vesicle from its outer space and has a low permeability with respect to hydrogen ions.

2) Electron transfer between ETC components along the membrane is accompanied by the transfer of protons across the membrane. This is realized owing to the asymmetric arrangement of carriers relative to both sides of the membrane. Several reasonable mechanisms of proton translocation were proposed (I shall not discuss them here). In chloroplasts, the translocation of protons and their subsequent partial exchange for other ions (as well as direct transfer of electron charges across the membrane) lead to acidification and positive charging of the thylakoid inner space relative to its outer space. The pH difference (ΔpH) and membrane electrical potential ($\Delta\psi$) are thus formed. In mitochondria, ΔpH and $\Delta\psi$ vectors have opposite directions (the outer space becomes more acidic and more positive). The resulting electrochemical potential (protonmotive force)

$$\Delta\mu_{H^+} = |\Delta\psi| + \frac{2.3\ RT}{F}\ |\Delta pH| \qquad (3.5)$$

is essentially the primary macroerg. Its discharge ensures ATP synthesis. It must be emphasized that in the realm of the "classical" chemiosmotic concept, $\Delta\psi$ and ΔpH are not local but bulk system characteristics.

3) $\Delta\mu_{H^+}$ synthesizes ATP by means of the membrane-bound enzyme, a ATP-synthetase, whose action is strictly vectorial. This enzyme forms a channel through which protons can permeate the membrane. As a proton passes through the ATP-synthetase channel towards low μ_{H^+} values (not necessarily one and the same proton), the system free energy decreases by $\Delta\mu_{H^+}$. This energy is utilized for ATP synthesis irrespective of the relative contributions of ΔpH and $\Delta\psi$ to the overall $\Delta\mu_{H^+}$ value. Formation of one ATP molecule requires the passage of several protons through the channel

of ATP-synthetase (their number is determined by the ratio of phosphate potential to proton-motive force). The enzyme works reversibly. In ATP hydrolysis, protons are transferred through the membrane backwards, and the membrane becomes "energized".

It is easy to see that Mitchell's hypothesis overcomes two objections that have been raised against the chemical concept. Firstly, the chemiosmotic concept can do without the chemical intermediates as yet not found. Secondly, it does not need large enough redox potential differences between neighboring carriers at coupling sites: energy contributions from all the coupling sites are summed up in the common pool.

WITT, in whose laboratory the most intrincate study of electron transport and phosphorylation kinetics in photosynthetic systems has been carried out, in his analysis of the bioenergetics of photosynthesis places strong emphasis on the first term in (3.5) [3.23]. According to Witt, the primary macroerg in the photophosphorylation process is the membrane electrical potential $\Delta\psi$, although in a steady state the concentration (ΔpH) and the electrical ($\Delta\psi$) terms are interchangeable.

SKULACHEV [3.24,25] also attaches great importance to the membrane electrical potential. Accordingly a cell possesses two forms of unified energy: chemical (ATP) and electrical ($\Delta\psi$). Energy transfer between individual intracellular structures is realized directly in the form of membrane potential. Energy liberated in the course of the functioning of any enzyme complex can be instantly utilized in other parts of a given mitochondrion, and through membrane reticulum in other mitochondria and other intracellular particles.

KELL [3.26] has recently subjected the chemiosmotic concept to fundamental revision. His own experiments, as well as the minute analysis of the published data, convinced him that $\Delta\mu_{H^+}$ values obtained by measuring "bulk" ΔpH and $\Delta\psi$ are obviously not large enough to be able to ensure ATP synthesis in the course of membrane phosphorylation (Sect.3.4). According to the orthodox chemiosmotic concept, during phosphorylation a membrane plays only the role of a diffusion barrier, but does not itself participate in energy accumulation. Kell's model likens a membrane to a fuel cell, whose functioning requires that the electrodes should be not at equilibrium with the surroundings. For exhaustive description of electrode processes one must take into account the presence of substantial potential barriers on the border between the electrode and the solution.

Kell's model is based on the following postulates:

1) The connection between electron transport and phosphorylation is ensured by the rise of a certain $\Delta\mu_{H^+}$ whose steady value is significantly higher than that measured between the bulk phases on both sides of a membrane.

2) A considerable part of the proton current during membrane phosphorylation does not involve the bulk water phases but is localized near the membrane surface within the so-called S-phases. Conductance is actualized along the chains of adsorbed water molecules according to Grotgus mechanism. The $\Delta\mu_{H^+}$ value at the ATP

synthetase inside the membrane is, therefore, not equal to $\Delta\mu_{H^+}$ between the bulk water phases.

3) An inalienable factor of membrane phosphorylation is the generation of a surface charge. This generation is actualized by the changes in the ionization constants of acid-base groups of ETC carriers and ATP-synthetase, and/or by the changes in their conformational states. The term "surface" stands for the near-membrane layer, the interphase S-phase, inside the Stern-Grachem layer [3.27].

4) The membrane is not just a diffusion barrier for protons but plays a role of an energy accumulator. This is caused by the existence of barriers which prevent the passage of protons through the interphase at the membrane-solution boundary.

It should be noted that the importance of surface charge generation was assumed before [3.28], but, at variance with Kell, the overall $\Delta\mu_{H^+}$ values at the near-surface layers and within bulk phases were considered to be equal.

Concurrently with Mitchell, another concept of membrane phosphorylation based on proton utilization was suggested by WILLIAMS [3.29]. A detailed account of WILLIAMS' views can be found in his reviews [3.30-33], according to which the most important events during membrane phosphorylation take place inside the membrane. The initial energy keepers (in a certain sense the primary macroergs) are ETC proteins-electron carriers transformed into their reduced forms by energy-donating redox reactions. The carrier reduction is accompanied by a change in proton activity within the membrane hydrophobic region resulting from the shift of redox equilibrium in the reaction of the $XH_2 \rightleftarrows X + 2H^+ + 2e^-$ type. In its turn, the change of proton activity in the hydrophobic phase can significantly shift the $H^+ + H_2O \rightleftarrows H_3O^+$ equilibrium and lead to considerable changes in the activity of water molecules. This in its turn, must lead to the ATP synthesis shifting the equilibrium of phosphorylation reaction. Naturally, all the aforesaid can take place only within a hydrophobic phase that does not contain an immence surplus of water molecules. The importance of protons is due to the fact that they are the sole movable metal ions capable of undergoing redox transformations. The intermediate macroergs directly inducing ATP synthesis are thus protons located within the membrane hydrophobic regions, their increased *local* chemical potential being out of equilibrium relative to the μ_{H^+} at near-the-surface membrane layers or within the bulk water phase. It is clear, therefore, that in the realm of Williams' concept, transmembrane gradients of ψ and pH do not participate directly in energy accumulation, storage, and transduction during membrane phosphorylation. It is incorrect, therefore, to regard as similar the concepts of Williams and Mitchell, designating the first one as "a microscopic chemiosmotic mechanism" [3.34].

Although in Williams' scheme a detailed physical analysis of the individual stages is lacking, this hypothesis signifies, from my point of view, an essential advance toward the understanding of the membrane phosphorylation mechanism. Most important is the notion concerning the decicive role played by the local chemical potentials and by the conformational changes of ETC carriers. The clarity of Williams' formulations in the analysis of the distinctions between his views and those of Mitchell, Witt and Skulachev is quite impressive [3.30,31].

Let us discuss now other concepts based on the conformational changes of proteins and membranes. The idea, according to which the conformationally changed and mechanically constrained states of macromolecules and their complexes play an essential part of bioenergetic processes, is very old. It goes back to the extremely interesting but, unfortunately, little-known works of BAUER [3.35]. The idea of energy accumulation and storage in the form of protein mechanical deformations during enzymatic catalysis was developed in [3.36-38]. These works were throughly discussed in monograph [3.2]. I shall briefly describe here the notions underlying the three best elaborated conformational concepts: the electromechanical model of Green, the unitary hypothesis of Bennun, and the conformational concept of Boyer.

It is appropriate, however, to begin with the work of PACKER [3.39] who proposed a conformational version of the chemiosmotic concept. Accordingly, the switching on of electron transport or the membrane energization during ATP hydrolysis leads, as in Mitchell's hypothesis, to a rise in ΔpH. The pH change induces changes in membrane protein protonation, which, in its turn, results in a conformational rearrangement of membrane proteins, and, through the lipid rearrangement, to changes in the ATP-synthetase structure able to ensure the ATP formation.

GREEN's electromechanical model [3.40,41] postulates the existence of a functional (and, maybe, a morphological) unit, a "supermolecule" that includes electron-transport complexes and ATP-synthetase. The substrate oxidation in mitochondria leads to the separation of electrons and protons in electron-transport complexes. These complexes become polarized, i.e., there appears an electric field, directed perpendicularly to the inner motochondrial membrane surface. This field in its turn induces a conformational transition in F_1 accompanied by the appearance of an electric dipole moment in the same direction (along the normal to the membrane surface). The whole supermolecule thus becomes polarized (energized). The phosphorylation substrates ADP and P_i enter the membrane in neutral, totally protonated forms. Polarization of the supermolecule leads to the appearance of charges on the molecules of phosphorylation substrates. The high-energy phosphate-ester bond is formed *in the course* of supermolecule depolarization. This is, probably, the most essential aspect of Green's scheme. The ATP synthesis acquires a kinetic character, it takes place during the transition from one stationary state to another. According to Green, the polarized ATP-synthetase retards electron motion through a coupling site. It lets the electrons pass again after depolarization and ATP synthesis. This explains naturally the respiratory control effect.

Discussing the physical aspects of bioenergetics, GREEN [3.42] has advanced an opinion that in the end there are two problems to be solved. The first one is energy transfer from the oscillator in the energy-donating center of the supermolecule to the oscillator in its energy-accepting center. According to Green, the essence of this problem is the fast and coherent transfer of vibrational excitation along, e.g., the "channels" of α-helixes (with the velocity of sound) and thus avoiding the thermal dissipation of the energy transferred. The second problem is the stability of excited states of the aforementioned oscillators. This problem can be reduced to the question of the appropriate structural organization. Green suggested concrete schemes of the electrochemical model realization in various bioenergetics processes.

The unitary bioenergetic concept developed by BENNUN [3.43-45] attaches an even greater importance to protein conformational changes. This hypothesis is based on the idea according to which energy transduction is realized by means of dynamic structural changes in macromolecules. Structural rigidity of a protein globule induces certain physical-chemical constraints on the outer regions of a macromolecule and its inner regions can, therefore, undergo vectored changes under the action of a scalar force. This leads to a nonuniform distribution of kinetic energy among movable parts of a macromolecule. In the energy-transducing systems the directional energy transfer from the energy-donating to the energy-accepting center is sustained by the protein synchronous conformational changes, in the course of which the energy-accepting center (ATP synthetase) undergoes changes lowering the activation energy of the energy-accepting process.

An interesting version of the conformational concept was suggested in BOYER's group [3.46,47]. ATP synthetase forms complexes with ATP, ADP and P_i, whose stability depends on the membrane conformational state. In an energized state, in which ADP and P_i are tightly bound to the protein, ATP binding is rather weak. In this state weakly bound ATP dissociates, and tightly bound ADP and P_i form ATP practically without any energy expenditure. The enzyme conformation changes (de-energization), the ATP formed becomes tightly bound, and the newly added ADP and P_i, weakly bound. Membrane energization induced by electron transfer leads once more to energized enzyme conformation, and the process is repeated all over again. The energy stored in the conformationally changed states of a membrane is thus mainly spent on two processes: 1) facilitation of P_i and ADP binding in a position optimal for ATP synthesis (according to Boyer, the most important is Pi binding), and 2) liberation of an ATP molecule tightly bound at the site of its formation.

Boyer's concept is schematically shown in Fig.3.1. In this scheme the rectangles represent a membrane bound ATP synthetase (□ conformationally equilibrium, and ⊏ conformationally nonequilibrium forms). Weak bonds with ligands are denoted as "-", and strong ones as "=". Under normal functioning, the membrane energization is actualized by redox transformations at ETC coupling sites. It is postulated that during the reverse process (ATP surplus, enzyme is functioning as an ATPase) all the steps are passed in a reverse sequence.

The amount of experimental data obtained in laboratories throughout the world during the years of membrane phosphorylation research is truly immense. The most

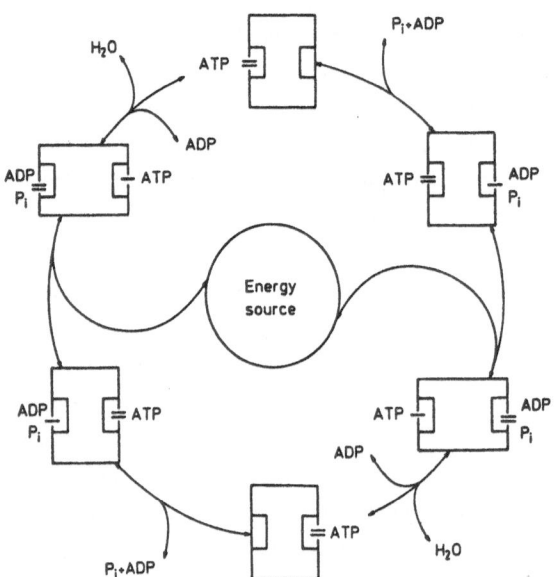

Fig.3.1. Boyer's scheme of ATP synthesis

convenient way to systematize and analyze these data is to discuss them from the point of view of one of the existing theoretical concepts (not necessarily the true one). It is appropriate to choose for this purpose Mitchell's hypothesis, which during the last 15 years has stimulated the majority of experimental studies in this field.

Before we pass on to the analysis of experimental data pro and contra Mitchell's concept, it will be helpful to consider two more general questions to clarify the meaning of two notions, which are used by everybody working in the field of bio-energetics. It will be done in the next two sections.

3.2 What is the Meaning of the Words: "Energy Coupling of Chemical Reactions"?

Almost any chemical reaction can be formally written as a sequence of two stages, one of which is energy-accepting, and the other energy-donating. For example, a well-known elementary reaction

$$H_2 + I_2 \rightarrow 2HI \tag{3.6}$$

can formerly be written in the form

$$\left. \begin{array}{l} H_2 \rightarrow 2H \\ I_2 \rightarrow 2I \end{array} \right\} \tag{3.6a}$$

$$I + H \rightarrow HI \quad . \tag{3.6b}$$

This division of reaction (3.6) into energy-accepting (3.6a) and energy-donating (3.6b) stages is, however, physically meaningless: both stages constitute one elementary act, or, using the terminology of the absolute reaction rate theory [3.48], proceed simultaneously through one and the same activated state. What is the meaning of the words: "simultaneously", "through one and the same elementary act"? Their meaning is obvious. The time interval during which events (3.6a,b) (i.e., the corresponding changes in interatomic distances and other system parameters) take place are so close to one another, or even overlap one another, that between them the absorbed or liberated energy does not have enough time to be distributed among the other statistical, vibrational, rotational, translational degrees of freedom, i.e., to be dissipated. There is no question of energy coupling of two reactions because there are no two reactions.

During the first years after the appearance of the chain reaction theory the notions of the so-called energy chains [3.49] were popular. According to these notions, the products of one elementary act possessing surplus energy ("hot" molecules and atoms) can take part in other elementary acts before their energy is dissipated due to collisions. The existence of energy chains in the case of certain reactions in gaseous phase has been shown recently [3.50]. Certainly, this type of energy coupling can be effective enough only in the case of gas reactions (and even then not very often). There are only two types of energy coupling mechanism in a condensed phase in chemistry. The first one was proposed more than 75 years ago by SCHILOV [3.51]. In this case, energy-donating and energy-accepting reactions have a common intermediary product, and coupling is realized in accordance with the mass action law.

As a simple example, let us consider consecutive reactions [3.52]:

$$A \underset{\longleftarrow}{\overset{K_1}{\rightleftharpoons}} B \qquad (3.7a)$$

$$B \underset{\longleftarrow}{\overset{K_2}{\rightleftharpoons}} C \quad . \qquad (3.7b)$$

Reaction (3.7a) is energy donating ($\Delta G^\circ < 0$, $K_1 > 1$) and reaction (3.7b) is energy accepting ($\Delta G^\circ_2 > 0$, $K_2 < 1$). We also assume that reaction (3.7b) was initially at equilibrium. Let us start now reaction (3.7a). This can be done by adding reagent A to the (3.7a) system if it is at chemical equilibrium, or by adding a catalyst to the system if it is shifted to the left of the equilibrium, but the (3.7a) reaction cannot proceed in the absence of a catalyst. The process will go on until both reactions reach the new states of chemical equilibrium.

What will then be the ratio between the number of acts of the energy-accepting reaction γ (i.e., the number of C molecules formed) and the number of acts of the energy-donating reaction α (i.e., the number of A molecules that have disappeared)? Let us designate this ratio as the process efficiency and denote it as β.

It is easy to show that after the new equilibrium has been reached

$$K_2 = \frac{C_0}{B_0} = \frac{C_0 + (\gamma/V)}{B_0 + (\alpha - \gamma)/V} \quad ,$$

where C_0 and B_0 are initial equilibrium concentrations of the corresponding reagent and V is the system volume. After simple transformations we get

$$\beta = \frac{\gamma}{\alpha} = \frac{K_2}{K_2 + 1} \quad .$$ (3.8)

It is important to note that β is determined only by the equilibrium constant of the energy-accepting reaction and depends neither on the initial state of the system nor on the energy-donating reaction parameters. The process efficiency approaches the maximum possible value, 1, when $K_2 \to \infty$. It means that one obtains a near to unity yield of the energy-donating reaction only if the energy-accepting reaction is energetically favorable [equilibrium (3.7b) is strongly shifted to the right], which contradicts our initial assumption.

It is easy to understand that this result is a corollary of two postulates: 1) both reactions are completely reversible; and 2) the reaction system is a thermodynamically closed system (after the addition of A), and reaction (3.7b) eventually reaches a new state of chemical equilibrium. If the process takes place in an open system (A is continuously added, and C continuously removed), the system can reach a steady state. In this state the concentration of intermediate B is constant, and, consequently, the number of eliminated A molecules is equal to the number of C molecules formed (per unit time), and the process efficiency

$$\beta = \gamma/\alpha = 1 \quad .[1]$$

As a matter of fact, the steady state is reached when the concentration of B becomes so large that in accordance with the mass action law the rate of B formation in the course of reversible reaction (3.7a) increases, and the rate of C disappearance in the course of reversible reaction (3.7b) decreases so much that these rates become equal. Thermodynamically unfavorable energy-accepting process is compensated by a superequilibrium concentration of the intermediate. At the beginning of the process before a steady state is reached, the yield of the energy-accepting reaction (i.e., the efficiency) will essentially depend upon the initial ratio of reagent concentrations.

These conclusions can be easily generalized and applied to systems of greater complexity comprising several consecutive reactions and involving not only monomolecular stages. Strictly speaking this first type of energy coupling mechanism does not represent energy coupling in the true sense of the word. There is no energy transfer between the acts of "energy-donating" and "energy-accepting" reactions. Energy

---

1 The β value is not a measure of the energetic efficiency of the process and is not connected with its performance coefficient. In a steady state the latter cannot reach 100%, part of the energy is obligatorily dissipated, and this dissipation is enhanced with the growing rate of the end product formation.

liberated in every act of the energy-donating reaction (if this reaction is exo-thermic) dissipate into heat, and energy absorbed in every act of the energy-accept-ing reaction (if this reaction is endothermic) is derived from a thermostat. In purely statistical systems only this type of coupling is possible.

Mechanisms of the second type require ordered constructions-machines. In ma-chines, energy can be localized on specific degrees of freedom which exchange ex-tremely slowly with thermal ones. Only under this condition can the energy liberated in an act of the energy-donating reaction ensure an act of the energy-accepting re-action before dissipation.

Two types of energy-transducing constructions-machines exist. They can be re-lated to two terms in Gibbs' expression for free energy, i.e., for that part of the energy of a system which can be utilized to perform external work:

$$\Delta G = \Delta H - T\Delta S \quad . \tag{3.9}$$

It is convenient to designate these two limiting cases as enthalpy and entropy machines. Let us consider the physical principles of the functioning of entropy machines.

During the work of any entropy machine two processes are realized: a thermodyna-mically favorable process, in which spontaneous decrease of the system free energy leads to an increase in its entropy without any heat effect, and a thermodynami-cally unfavorable process of the transformation of heat into useful work. Entropy machines function owing to ambient heat, to a thermostat. Only two kinds of entropy machines exist.

It is appropriate to analyze the functioning of entropy machines of the first kind using as an example the expansion of ideal gas into vacuum. A cylinder subdi-vided into two volumes V_1 and V_2 by a hermetic partition is placed in a thermostat at temperature T. The partition can be moved out, friction is neglected. Volume V_1 initially contains a gas under pressure P_1 at thermal equilibrium. The equation of state holds true

$$pV_1 = nRT \quad , \tag{3.10}$$

where n is the number of gas moles. Volume V_2 is vacuous. Let us move the partition out, i.e., let the gas volume increase to $V_1 + V_2$. It is an isothermal process without any thermal exchange with the thermostat (the gas is ideal). The gas pres-sure now equals $P_2 = P_1V_1/(V_1 + V_2)$. The intrinsic energy of the system does not change, and the entropy increases by $nR \ln(V_1 + V_2)/V_1$. The free energy of the sys-tem was decreased by $nRT \ln(V_1 + V_2)/V_1$, but it has wholly turned into bound ener-gy, no useful work has been done.

Let us now change the design of the device and carry out isothermal expansion in a different way (Fig.3.2). There is now a piston instead of a partition. This piston is assumed to be weightless, its friction with a wall is neglected. There is a weight M_1 on the piston, so that

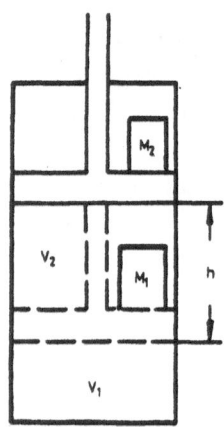

Fig.3.2. Device for performing external work using isothermal expansion of the ideal gas

$$gM_1/A = P_1 = nRT/V_1 \quad , \tag{3.11}$$

where g is the acceleration of gravity, A is the piston area. Let us exchange (instantly) M_1 for a smaller weight M_2, so that

$$gM_2/A = P_2 = nRT/(V_1 + V_2) \quad . \tag{3.12}$$

The gas expands isothermally (now due to thermal exchange with the thermostat),
and the piston will ultimately get into a new equilibrium position. In this new
state the gas will again have temperature T_1, volume $V_1 + V_2$, and pressure P_2.

If we do not include M_1 and M_2 in the system, we can say that the system's free
energy, as in the first case, has been decreased by $nRT \ln(V_1 + V_2)/V_1$, entropy
has been increased by $nR \ln(V_1 + V_2)/V_1$, and bound energy increased by
$nRT \ln(V_2 + V_1)/V_1$. Our system has also performed external work M_2gh. The heat
$Q = M_2gh$ obtained by the system from the thermostat has been used to perform this
work. It is easy to show that in the case of infinitely slow weight decrease (from
M_1 to M_2) the limit value of the heat received and the work performed will be equal
to $nRT \ln(M_1/M_2) = nRT \ln(V_1 + V_2)/V_1$, i.e., exactly equal to the free energy decrease in the gas.

What is the mechanism of the transformation of heat into work in this entropy
machine? The specific degree of freedom corresponds to the mechanical motion of the
piston along the cylinder axis. The excitation of this degree of freedom is realized
by elastic collisions between gas molecules and piston surface as a result of which
the molecule transfer a part (in accordance with the law of pulse conservation) of
their kinetic energy to the piston with a weight. Excitation is initiated by the
fast replacement of wheights. The possibility of this excitation is determined by
the machine construction (cylinder with a piston) and by the fact that the piston
and the weight are solid bodies with rigid bonds between their atoms.

In the entropy machines of the first kind the heat to work transduction is thus
actualized with the help of a device able to excite the *mechanical motion* of the
parts of the construction by utilizing the *kinetic energy* of molecules.

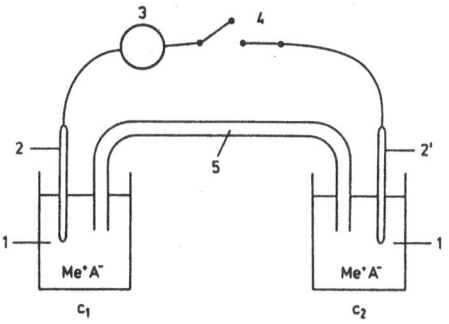

Fig.3.3. Concentration galvanic cell (1: half-cells; 2 and 2': Me electrodes; 3: electrolyzer; 4: key; 5: liquid junction; Me⁺A⁻: strong electrolyte; c_1 and c_2: molar concentrations

Let us now turn to the entropy machines of the second kind. The principles of their functioning can be illustrated using, as an example, a concentration galvanic cell-electrolyzer system. Figure 3.3 shows the scheme of this system. Molar concentration of a strong electrolyte MeA (for simplicity the electrolyte ions are assumed to be singly charged, Me^+, A^-) in the left half-cell is c_1, and in the right half-cell, $c_2(c_1 > c_2)$, the solution volumes in the half-cells are the same and equal to V. The solutions are assumed to be ideal. Without closing the circuit, let the electrolyte be diffused from the left half-cell into the right one up to the equalization of concentrations. It is easy to show that if we neglect small volume changes, as a result of the concentration equilibrium being established the system's entropy increases by

$$\Delta S = RV\left(c_1 \ln \frac{c_1 + c_2}{2c_1} + c_2 \ln \frac{c_1 + c_2}{2c_2}\right) . \tag{3.13}$$

Free energy decreases by

$$\Delta G = RTV\left(c_1 \ln \frac{c_1 + c_2}{2c_1} + c_2 \ln \frac{c_1 + c_2}{2c_2}\right) . \tag{3.14}$$

and bound energy increases by the same amount. In this case the establishment of concentration equilibrium leads to the transfer of a part of the system total energy from the free into the bound form with no useful work performed.

Let us reach the concentration equilibrium in a different way using the electric potential difference on the electrodes of the concentration cell, which is equal to

$$Z = \frac{RT}{F} \ln \frac{c_1}{c_2} . \tag{3.15}$$

The liquid junction 5 ensures the absence of the diffusion potentials. Let us close key 4. Concentration equalization will now proceed only through electrode processes, the equilibrium of which is shifted in such a way that overall electrode reactions proceed on electrodes 2 and 2':

$$Me^+ + e(Me) \rightarrow Me \text{ (electrode 2)} \tag{3.16a}$$

$$Me \rightarrow Me^+ + e(Me) \text{ (electrode 2')} . \tag{3.16b}$$

43

The external circuit current passes through an electrolyzer performing useful work (decomposing water into H_2 and O_2). What is the maximum useful work which can be ensured by a concentration cell? Let the cell be completely discharged through the electrolyzer. Electrolyte concentrations in the half-cells become equal, the weight of electrode 2' decreases and that of electrode 2 increases by the same value equal to the weight of the Me^+ ions transferred (this weight can be calculated taking into account that concentration changes in each half-cell are equal to $[(c_1 + c_2)/2]$. The resultant transfer of neutral metal atoms from the metal lattice of electrode 2' to the metal lattice of electrode 2 is accompanied neither by energy nor entropy changes. An increase in the system's entropy due to the equilibration can be thus estimated, as in the first case, with the help of (3.13), and a decrease in the system's free energy (equal to the increase in the system's bound energy) with the help of (3.14). External work A can be estimated with the help of (3.15), taking into account the gradual decrease of potential difference while equilibrium is being approached. The maximum useful work can be performed in the case of an infinitely slow process and, with Joule losses neglected, is equal to

$$A = RTV\left(c_1 \ln \frac{c_1 + c_2}{2c_1} + c_2 \ln \frac{c_1 + c_2}{2c_2}\right) \quad , \tag{3.17}$$

i.e., precisely equal to the decrease in the system's free energy resulting from the equalization of concentrations. As in the case of any entropy machine this work is performed owing to the heat. The mechanism of transduction from the thermal into the electric form of energy of separated charges, incapable of exchanging directly with statistical degrees of freedom, is quite evident. Elementary acts (3.16a) and (3.16b) increase the potential energy of two corresponding Me^+ ions by $\Delta Z.e$ [ΔZ is determined according to (3.15)]. The charge particle must in every case overcome a potential barrier on the electrode-solution boundary. This boundary functions as a Maxwell's demon, selecting the "hot" ions capable of overcoming the potential barrier, absorbing the heat from the environment and thus maintaining ΔZ.

In this device the energy-accepting reaction (electrochemical decomposition of water) is spatially separated from the energy-donating reaction with the galvanic cell. Because of the specificity of construction (proper connections of wires, electrodes, etc.) every act of energy-donating reaction (electrochemical transformation of ions on the electrodes of the concentration cell) is, however, accompanied, in the average, by one act of energy-accepting reaction (discharge of H_3O^+ and OH^- ions on the electrodes of the electrolyzer). In this case we can consider conditionally the energy-donating and energy-accepting reactions to be taking place in the course of one elementary act: the energy liberated as a result of an energy-donating process (more accurately, the energy obtained from the thermostat with the help of an energy-donating electrochemical process) excites a specific electrical degree of freedom and is transferred without dissipation (neglecting Joule's losses) to ensure an energy-accepting process.

As a matter of fact, the principles underlying the functioning of entropy machines of this kind are not different from the aforementioned Shilov's coupling principle through a common intermediate. In the concentration galvanic cell-electrolyzer system the role of the common intermediate is played by electrons. One of the functions of the device is, in this case, to decrease sharply the system's capacity. A steady state with maximal for this process of the intermediate concentration is reached almost instantly, and after that every act of energy-donating process is, in the average, accompanied by one act of energy-accepting process. Scheme (3.7) with two consecutive reaction is, in the final analysis, also a primitive entropy machine. The role of constraint is played, in this case, by the walls of reaction vessel, which fix the system's volume and allow the concentration of intermediate product B to be increased. Evidently, in the case of any mechanism based on the mass action law, nothing will be altered if we take into account the membranes and the osmotic phenomena [3.53,54].

In the afore-described entropy machines and chemical systems of the Shilov' type the energy-accepting processes are realized via the energy obtained from a thermostat and not directly from the energy-donating processes. If two reactions are coupled as in the system (3.7), in accordance with the mass-action law, the "act per act" coupling in the steady state can be actualized only on the average.

Only when the energy coupling of chemical reactions is realized by means of a machine, which excites specific degrees of freedom not involving a thermostat and ensures direct energy transfer without dissipation (except friction losses) between the energy-donating and the energy-accepting processes, is it possible to have true act-per-act type of coupling even before a steady state is reached. This provides us with the most straightforward way to check experimentally the physical mechanism of energy coupling in a biological system.

We have described here two types of entropy machines differing from one another by the mechanism of elementary acts transforming heat into a "noble" form of energy. In machines of the first type this mechanism consists in the transfer of mechanical moments of particles colliding with the rigid surface of a macroscopic body (a piston, in this case) and, owing to the specific construction of the device, exciting a mechanical translational degree of freedom. In machines of the second type, this mechanism is reduced to increasing the potential energy of the particles involved in elementary acts resulting in the "selection" of the "hot" particles from a reaction mixture, charge separation and a rise in potential difference not being obligatory.

Both these examples of entropy machines are far removed from biology. They were chosen for clarity and simplicity, as well as because they exhaust all the possible mechanisms of entropy machine functioning.

3.3 Transmembrane Electrochemical Potential, its Components and Physical Principles of its Utilization in Bioenergetic Processes

According to the basic postulate of the chemiosmotic concept, the source of energy required for ATP synthesis during membrane phosphorylation process is $\Delta\mu_{H^+}$, the transmembrane difference of the hydrogen ions' electrochemical potentials on the both sides of a membrane. This difference can be presented in the form of a sum of two separately measurable items: transmembrane electric potential $\Delta\psi$, and the difference of the chemical potentials of hydrogen ions on both sides of a membrane, determined by the difference in their activities (approximately concentrations) (3.5). Electrochemical potential as a thermodynamic characteristic of a system containing charged particles has been introduced by Guggenheim. An exhaustive physical analysis of this quantity and its components can be found in [3.55].

The value of the electrochemical potential $\bar{\mu}_i^\alpha$ of ion i in a certain phase α depends not only on the phase composition but on its "electrical state" as well. The "electrical state" of the phase depends, naturally, on its composition because electrical charges do not exist without their carriers — electrons, cations, and anions. The electric charge of an ion is, however, so large that the charging of any macroscopic phase up to the highest possible potential requires a practically negligible change in its composition. This allows us to consider the two phases to have the same chemical composition but different electrical states. The difference between the electrochemical potentials of ion i in phases α and β will be determined, in this case, by the difference between electric potentials of these phases and by the charge of the ion z_i (in proton charge units):

$$\bar{\mu}_i^\beta - \bar{\mu}_i^\alpha = \mathcal{F}z_i(\psi^\beta - \psi^\alpha) \quad , \tag{3.18a}$$

where $\mathcal{F}$ is faraday. For two phases of the same composition the difference between their electric potentials can be thus determined thermodynamically:

$$\psi^\beta - \psi^\alpha = \frac{1}{z_i\mathcal{F}}(\bar{\mu}_i^\beta - \bar{\mu}_i^\alpha) \quad . \tag{3.18b}$$

For phases with different compositions one can formally write:

$$\bar{\mu}_i^\beta - \bar{\mu}_i^\alpha = (\mu_i^\beta + z_i\mathcal{F}\psi^\beta) - (\mu_i^\alpha + z_i\mathcal{F}\psi^\alpha) \quad , \tag{3.19}$$

where μ_i^β and μ_i^α are "purely chemical" potentials of ion i in phases β and α respectively. For most real systems such division of $\bar{\mu}_i$ into electrical and chemical components is, however, physically meaningless. In the case of phases having different chemical compositions such division can be well-grounded only for ideal (infinitely diluted) solutions of ions in totally identical solvents in both phases. Then

$$\psi^\beta - \psi^\alpha = \left(\bar{\mu}_i^\beta - \bar{\mu}_i^\alpha - RT \ln \frac{N_i^\beta}{N_i^\alpha}\right) \quad , \tag{3.20}$$

where N_i^α and N_i^β are the molar fractions of ion i in phases α and β, respectively. The individual terms in (3.19) have a physical meaning only under the aforesaid limiting conditions. These conditions are, naturally, never met in real membrane systems. In heterogeneous biological systems one can, as a rule, speak of a difference between electrochemical potentials of an ion only in the sense of a change in the system free energy in the course of a certain chemical reaction. To be sure, it is possible to evaluate this energy change in volts using well-known reduction coefficients. The physical meaning of the values obtained is, in this case, quite clear. Such electric potential difference could be obtained if we utilized for this purpose the free energy decrease in the course of a given chemical reaction with the help of *a device specific for a given ion.*

Strictly speaking it can be stated that a certain part of the electrochemical potential difference exists in the form of a real macroscopic electric field only under the aforementioned limiting ideal conditions. We must, therefore, be extremely careful with the so-called measurements of transmembrane electric potentials in biological systems and subject the measurement procedure to painstaking analysis. Similar observations have been made by LOWENHAUPT [3.56]. After these to no small degree trivial general remarks it is appropriate to return again to the principal question concerning the possibility of utilizing the electrochemical potential differences of hydrogen ions in bioenergetic processes.

Let us now forget that subdivision of $\Delta\mu_{H^+}$ into the electric and concentration components lacks, as a rule, any clear physical meaning. Let us imagine a completely impenetrable (for protons) chemically neutral partition separating two identical water solutions differing from one another only by proton concentrations, and a properly directed *external* electric field ensuring the electric potential difference on both sides of this partition. A channel for protons exists which is connected with a certain device able to perform work (e.g., to synthesize ATP) by utilizing the energy liberated in the course of proton transfer through this channel along the gradient of the proton electrochemical potential. What requirements must this device meet in order to be able to utilize interchangeably both components of $\Delta\mu_{H^+}$ at its disposal?

To be sure, we may confine ourselves to the thermodynamic possibility and do not concern ourselves with the physical mechanisms of energy transduction by elementary acts. In this case the reasoning is quite simple. There are two coupled thermodynamic flows: the flow of proton transfer and the flow of ATP synthesis. The more protons are transferred, the more ATP molecules are synthetized. One can increase the proton flow, i.e., the number of protons transferred through the synthetase channel per time unit, increasing either the proton-concentration difference at the two sides of the membrane, or the external electric field of the proper sign, as well. Both thermodynamic forces that ensure the increase of proton flow, i.e., grad(pH) and $\Delta\psi$ are naturally interchangeable. This obvious interchangeability, however, concerns only the maintenance of the proton flow through the membrane, but not the coupled flow of ATP formation. In order to provide for the interchangeability of this two forces relative to the ATP formation there must exist special mechanisms for utilizing the energies liberated due to ΔpH and $\Delta\psi$.

If the thermodynamic force connected with $\Delta\psi$ is used only to increase the proton flow, it means that the coupling is based on the Shilov type of mechanism. Proton (or hydroxyl ions) play the role of a common intermediate, and in steady state the concentration (but not the concentration gradient) of H^+ or OH^- must be high enough to ensure the stoichiometric (relative to electron transfer) ATP formation at a given value of phosphate potential, $P = (a_{ATP})/(a_{ADP} \cdot a_{P_i})$.

Until this steady state is reached, the coupling of the "act per act" type cannot be realized. We shall see below that experimental data do not satisfy the afore-listed requirements, and that any mechanism of the Shilov type must be rejected.

An alternative mechanism makes use of the ΔpH by means of an entropy machine, and $\Delta\psi$ increases not the proton flow but the energy utilized for the ATP synthesis in the course of every elementary act.

It is thus necessary for our machine to be able to function using ΔpH and $\Delta\psi$ either separately or together to perform one and the same process.

Let us begin with ΔpH. The system's free energy decrease due to proton transfer from the region of high proton concentration to the region of low proton concentration is determined exclusively by the entropy component of (3.9): the total energy of each proton in the absence of $\Delta\psi$ is the same on both sides of the partition. The ATP synthesizing device (let us designate it as ATP synthetase for the sake of simplicity) must then function as an entropy machine. In accordance with the preceding section, it means that ATP synthetase must either utilize the kinetic energy of the protons transferred (an entropy machine of the first kind), or possess chemical groups able to react with the protons binding them on one side of the partition and liberating them on the other, the products of these reactions having enhanced potential energies (an entropy machine of the second kind).

The heat which ensures this reversible process is derived from the thermostat. Naturally, both mechanisms must involve the excitation of slowly relaxing degrees of freedom: an elastic deformation of ATP synthetase or of the whole membrane, the formation of a more or less stable but thermodynamically unfavorable chemical bond, a rise in the electric potential difference, etc. I shall not discuss here the immediate mechanism of ATP synthesis, i.e., the mechanism whereby the already created "noble" form of energy is transformed into the energy of the ATP ester-phosphate bond. The difficulties arise even before that.

It has already been stressed that according to the chemiosmotic concept, the machine which synthetizes ATP in biological membranes must know how to utilize $\Delta\mu_{H^+}$ irrespective of the energy distribution between its electric and concentration components. We shall, therefore, pay special attention to the conditions of interchangeability of these components in our analysis of different mechanisms of $\Delta\mu_{H^+}$ utilization.

Let us firstly assume that ATP-synthetase utilizes ΔpH as an entropy machine of the first kind using proton kinetic energy. Taking this into account, it is then necessary that the electric field should act in such a way as to increase the proton kinetic energy, i.e., speed up the protons during their "flight" through the ATP-synthetase channel (otherwise it would be necessary to accept that ATP synthetase has two independently functioning devices, one of which synthesizes ATP utilizing the proton kinetic energy, and the other utilizing the energy conserved in $\Delta\psi$). In other words, a decrease in free energy of each proton that passes the electric potential difference, must, in this case, be transformed with high efficiency into kinetic energy of this proton:

$$\Delta\psi \cdot e \approx E_{kin} + \text{friction loss} \quad ,$$

where $E_{kin} = mv^2/2$.

In a condensed phase, however, such selective "heating" of protons (as well as of other particles) is impossible. It is well known that the lifetime of such "hot" particles in a condensed phase cannot be more than 10^{-12} s. An electric field cannot, therefore, influence the functioning of the entropy machines of the first kind, and these must, therefore, be left out of our consideration.

Entropy machines of the second kind whose functioning has been analyzed in detail above, using a concentration galvanic cell as an example, can, in their turn, be of two varieties. In the first one the elementary acts, in the course of which heat is transduced into a "noble" form, do not lead to charge separation and creation of electric potential difference. The energy derived from a thermostat in the course of every elementary act is determined by the type of reaction between ATP synthetase and hydrogen ions, and cannot depend on electric field. Devices which utilize ΔpH and $\Delta\psi$ must in this case function independently from one another, and these two energy sources cannot be interchangeable.

One last possibility remains. The pH-utilizing entropy machine transforms heat into the electrostatic energy of separated charges, i.e., creates an electric potential difference to be added to the membrane potential $\Delta\psi$. ATP synthetase (strictly speaking, the membrane region through which hydrogen ions are transferred from the high-into the low-concentration side) must in this case function as a concentration galvanic cell. Any other way of obtaining a potential difference as a result of the concentration difference would lead to the wrong sign of this additional $\Delta\psi$.

ATP synthetase contacts with water phases on both sides of the membrane must thus be essentially actualized by the hydrogen electrodes. An electric field of the same sign as that of the membrane potential rises up within the ATP-synthetase across the membrane. The decrease of the electrostatic energy of protons during their passage through the channel of ATP synthetase must ensure ATP synthesis. It cannot be done by increasing the kinetic energy of protons in the electric field:

in a condensed phase this will necessarily lead to energy dissipation. The sole possibility to "spend" this energy in a useful way is to enhance the potential energy of the particles taking part in the reaction by either increasing the distances between opposite charges or decreasing the distances between charges of the same sign. When one tries to invent a mechanism of this type it must be borne in mind that pH has already been utilized to create the additional electric field and cannot any longer serve as a free energy source.

We can, therefore, conclude that the only way to utilize interchangeably both components of $\Delta\mu_{H^+}$ in order to ensure a thermodynamically unfavorable process requires not only the fulfillment of hardly feasible conditions but is reduced to the obligatory generation of an additional electric potential by ΔpH. It means that if $\Delta\psi$ is measured properly, i.e., what is actually measured is indeed the electric field in the region of the energy-transducing device, one cannot sum up the measured values of $\Delta\psi$ and RT/F (ΔpH) (3.5) because ΔpH had been already taken into account in the measured $\Delta\psi$ value. According to my point of view this analysis shows that the chemiosmotic concept in its classic version (i.e., while this concept remains "chemiosmotic") cannot be regarded as a proper foundation of the membrane phosphorylation theory.

3.4 The Chemiosmotic Concept. Experimental Data Pro and Contra

In this section we shall discuss only the energy aspects of Mitchell's hypotheses, and even more specifically, Mitchell's principle of the coupling between energy-donating and energy-accepting reactions in membrane phosphorylation. Brilliant works and conjectures of Mitchell concerning the asymmetric arrangement of carriers in energy-transducing membranes and the resultant generation of concentration differences of protons and other ions on both sides of the membranes are major achievements of modern biochemistry.

Let us enumerate again the main propositions of the chemiosmotic concept which require experimental proofs.

1) Electron transfer along ETC leads to a rise in the proton electrochemical potential difference across the topologically closed energy-transducing membrane ($\Delta\mu_{H^+}$) whose value is high enough to ensure thermodynamically ATP synthesis under cell conditions.

2) Both components of $\Delta\mu_{H^+}$, i.e., the transmembrane electric potential ($\Delta\psi$) and the transmembrane proton concentration difference (ΔpH), are equally competent in ATP synthesis.

3) The "discharge" of $\Delta\mu_{H^+}$, i.e., the passage of protons down the μ_{H^+} gradient through the coupling factor (ATP synthetase), leads to ATP synthesis irrespective of the way $\Delta\mu_{H^+}$ generated.

Before discussing the experimental data concerning the Δψ, ΔpH and phosphate potential measurements in the course of oxidative and photosynthetic phosphorylation, it is necessary to dwell on the methods of their measurements, especially on the methods of Δψ measurements which usually entail arbitrary assumptions (to say nothing of the afore-mentioned difficulties connected with subdivision of $\Delta\mu_{H^+}$ into two independently measurable components).

3.4.1 Methods of Δψ Measurements

The most widespread method of Δψ measurement in mitochondria is based on estimating the distribution of penetrating ions between the matrix and the environment. The method implies the following picture of this process. Because of electron transport along mitochondrial ETC in the conditions of asymmetric distribution of carriers relative to the inner and the outer surfaces of mitochondrial inner membrane, as well as owing to various proton-donating and proton-accepting processes within the matrix and on the other side of the inner membrane, ΔpH rises: protons are pumped out of a mitochondrion inner space with its alkalization. There is no considerable passive proton transfer: the membrane is practically impenetrable for protons. As a result of unequal proton absorption on the inner and the outer membrane surfaces, substitution of protons by other cations, and of other processes not connected directly with proton transfer, there appears a membrane potential ("+" out, "-" in).

If a surplus of freely penetrating ions is now introduced into the system (e.g., K^+ ions in the presence of valinomycin), these ions will be unevenly distributed between both sides of the mitochondrial inner membrane.[2] It is postulated that this uneven distribution is caused by the field of membrane potential which exists *independently* of the added freely penetrating ions.

In the aforementioned example, one assumes that K^+ ions diffuse passively into the matrix in the electric field of membrane potential, and their inner concentration is raised until the rate of their backward diffusion, in accordance with the Fick law, becomes equal to their diffusion rate in the electric field. It will evidently happen when the ratio of K^+ concentrations inside ($[K^+]_i$) and outside ($[K^+]_o$) the matrix satisfies Nernst's equation:

$$\Delta\psi = \frac{RT}{F} \ln \frac{[K^+]_i}{[K^+]_o} \quad . \tag{3.21}$$

This was the method of Δψ measurement used by MITCHELL and MOYLE in 1969 [3.58]. Other authors applied the same method. NICHOLS [3.59] used a rubidium radioactive

---

2 Measurements of ion distribution are, as a rule, carried out using isolated mitochondria and estimating ion concentrations within the matrix and in the environment. It is accepted (and most probably quite reasonably [3.57]) that ions are distributed evenly between the environment and the space separating the mitochondrial inner and outer membranes.

isotope with valinomycin, KAMO et al. [3.60] a freely penetrating tetraphenylphos-
phonium cation. LIBERMAN, SKULACHEV and their co-workers have developed similar
methods of $\Delta\psi$ measurement based on the distribution of easily penetrating anions
[3.21,25,61-65].

It is, however, easy to make sure that the reasoning underlying these methods is
wrong. The uneven distribution of freely penetrating ions between the environment
and the matrix of functioning mitochondria in a steady state is, of course, an in-
disputable fact: the concentration of cations increases, and that of anions decreases within the matrix. This fact could, in principle, be explained assuming the
existence of specific pumps pushing forcibly, e.g., K^+ ions into the matrix in exchange for the protons pumped out of the matrix and thus preserving the electroneu-
trality of the system [3.66]. According to such mechanism the uneven distribution
of freely penetrating ions is maintained by the electroneutral pumps, and not by
$\Delta\psi$. This explanation, however, does not seem to be well grounded; it is rather difficult to imagine one pump realizing the active transport of various ions, those
alien to the cell included. It is even more difficult to visualize a set of different pumps stored up for all the types of penetrating ions that can be added by a
scientist to functioning mitochondria. It is thus evident that the uneven distribution of freely penetrating ions in steady-state mitochondria means an obligatory
existence of a transmembrane electric potential of the proper sign whose magnitude
gives (3.21). It does not mean, however, that this potential did actually exist
in the absence of the penetrating ions added by a researcher for $\Delta\psi$ measurements.

Let us assume that ETC functioning results in the formation of ΔpH across the
mitochondrial membrane, which is almost impenetrable for protons, but in the absence
of added freely penetrating ions is *not* accompanied by a rise in $\Delta\psi$. Let the amount
of protons translocated from the matrix after the beginning of the process be equal
to δ_{H^+}. The latter can be estimated if we know the values of pH_i and pH_o, the volumes
v_i and v_o, as well as buffer capacities P_i and P_o ($P = v^{-1} \cdot d\delta H^+/dpH$) of water
phases inside the matrix and outside mitochondria, respectively. The membrane being
practically impenetrable to protons, this system in a steady state is equivalent to
a system in a state of Donnan equilibrium (we must bear in mind that the mitochondrial
membrane is completely penetrable to water molecules and represents a perfect osmo-
meter [3.57]). The system in a steady state will, therefore, behave as though there
exists a nondiffusing anion X^- localized within the matrix with the concentration
of $[X^-]_i \approx \delta H^+/v_i$ [3.66]. Under these conditions a freely penetrating cation (e.g.,
K^+ in the presence of valinomycin) will be distributed between the matrix and the
environment according to Donnan's distribution law. The number of K^+ ions transferred
into the matrix will then be (if the amount of other easily penetrating ions is much
smaller than that of K^+ ions) approximately equal to the number of protons trans-
ferred to the outside up to the moment a steady state is reached ($\delta K^+_{o \to i} \approx \delta H^+_{i \to o}$), and
the concentration of K^+ ions within the matrix will be approximately equal to
$[X^-]_i$ ($[K^+]_i \approx H^+/v_i$).

Such distribution *will lead to the formation* of a membrane potential:

$$\Delta\psi = \frac{RT}{F} \ln \frac{\delta H^+}{v_i[K^+]_o} \quad .$$

(3.22)

(A splendid account of Donnan equilibrium theory can be found in [3.67]).

These measurements cannot, however, give any idea of what membrane potential value is equal to before the addition of freely penetrating ions (i.e., for instance, before the addition of valinomycin). The membrane potential owing to Donnan's distribution of rather small quantities of readily and poorly penetrating cell ions cannot be determined. It depends in a complicated manner on the rate of proton transfer, the diffusible ion mobility, the changes in intracellular effective water volume, etc. One can be sure that the true membrane potential is lower than that estimated by the above method. The membrane potential which actually rose independently of ΔpH formation and the resulting Donnan's distribution can be estimated rather roughly by subtracting $\Delta\psi$ value obtained with the help of (3.21) from the value obtained with the help of (3.22). In this way we estimate the contribution of the passive transfer of added penetrating ions in the field of initially existing (i.e., the electric field not created by Donnan's distribution of added penetrating ions) into the observed distribution of these ions.

Two other notions concerning the distribution of penetrating ions have to be touched upon. It has been already mentioned that isolated mitochondria are ideal osmometers. Transfer of penetrating ions into the matrix must, therefore, be accompanied by water transfer and by an increase in the matrix effective free water volume. Most authors do not, however, take this into account and estimate the concentration of penetrating ions by dividing their amount by the matrix free volume measured in the absence of the added penetrating ions [3.58]. $\Delta\psi$ values obtained with the help of (3.21) are, in this case, somewhat overestimated.

One can find statements in the current literature that the addition of penetrating ions (e.g., K^+ ions with valinomycin) destroys $\Delta\psi$. This is often used to check one of the propositions of the chemiosmotic concept, according to which $\Delta\psi$ existence is a compulsory condition of ATP synthesis [3.68]. Such statements are, obviously, incompatible with the application of penetrating ions for $\Delta\psi$ measurements. If added ions are unevenly distributed on both sides of the membrane, and their active transport with the help of a specific pump is excluded, there must then exist a properly directed $\Delta\psi$ in the presence of these ions. Elementary logic does not allow us to explain the uneven distribution by $\Delta\psi$, and the $\Delta\psi$ absence by the uneven distribution.

Another method used to estimate $\Delta\psi$ is based on measuring the intensity of the chlorophyll delayed red fluorescence in chloroplasts. Delayed fluorescence, the duration of which reaches milliseconds and even seconds, is caused by the recombination of photochemically separated electric charges in the active center of photosystem 2 [3.69]. In order to reverse one of the primary acts of the photosynthesis light stage, certain activation barriers must be overcome: the energy level of a system comprising a separated electron and a hole localized, respectively, on the electron-accepting and the electron-donating centers of photosystem 2 lies somewhat lower than the fluorescent singlet excited level of a chlorophyll molecule in the pigment antenna. Fluorescence intensity J is, therefore, proportional to

$$J \sim e^{-E_a/RT} \quad , \tag{3.23}$$

where E_a is the activation energy of the charge recombination process, The influence of an artifically created membrane potential on the delayed fluorescence has been shown for the first time in Barber's laboratory [3.70], where the corresponding method of $\Delta\psi$ measurement in photosynthetic objects has been also developed [3.71]. This method is based on the assumption that donor and acceptor centers carrying separated electric charges make contacts with the water phases on different sides of the thylakoidal membrane [3.72], so that $\Delta\psi$ raises the energy level of the separated charges and consequently lowers the activation energy of their recombination. It is, however, doubtful whether this method can be used to measure $\Delta\psi$ between water phases on different sides of a membrane. The question is, in fact, of the effect on the positions of the occupied electron level of the acceptor carrying a negative charge, and the empty electron level of the donor carrying a positive charge. The positions of these levels will be affected not only by the transmembrane electric potential but by any local changes in the charge distribution both on the membrane surface and within the membrane volume near the donor and acceptor centers. It is rather doubtful, therefore, that such measurements can serve as the basis of any quantitative and even qualitative conclusions concerning the values of transmembrane electric potentials in real energy-transducing systems whose functioning is accompanied by considerable structure and charge rearrangements.

A method, based on the so-called electrochromic displacement of the absorption bands of dyes either endogenous or artifically introduced into a membrane, has recently become very popular in photosynthesis studies mainly because of its intensive application in Witt's laboratory, which is one of the leading world centers in this field [3.23,73]. This method was first used in bioenergetic research by JACKSON and CROFTS who studied the energized states in chromatophores of certain photosynthetic bacteria [3.74], and in Witt's laboratory, in the chloroplasts of green plants [3.75-77]. The term "electrochromism" is applied to the spectral absorption changes due to perturbation and displacement of molecular electron levels in an electric field. The appearance of shifted bands in the absorption spectra of photosynthetic systems under illumination was discovered long ago [3.78,79]. It is clear now that these shifted bands in the ~520 nm wavelength region belong to carotenoids. The results with model systems (carotenoid films between the plates of a capacitor [3.80]) repeat qualitatively those obtained in vivo, and a calibration of the electric field by the intensity of a shifted band has been carried out in the assumption of

a linear dependence between the extinction changes and the electric field strength
[3.72,73]. Certain doubts concerning the adequacy of this method in measuring membrane electric potential have been recently expressed in literature. First of all,
sufficiently long illumination (of ~1 s duration and longer) changes the thylakoid
shapes which results in the changes of light-scattering, masking the electrochromic
shifts [3.72]. Not all is well, however, in the case of experiments using illumination of much shorter duration. A detailed study of the connection between the
"carotenoid shift" and ATP synthesis in chromatophores and chloroplasts has been
carried out recently [3.81,82]. It was shown that there indeed exists a correlation
between the carotenoid shift and the energized state formation, but (at least for
the chromatophores of Rhodospirillum rubrum) a rather small part of the changes observed, not exceeding 30%-40%, is associated with energization. The duration of
these slow changes (associated with energization) is characterized by $\mathcal{T}_{\frac{1}{2}} \approx 10$ ms.
Fast changes with $\mathcal{T}_{\frac{1}{2}} < 2$ μs do not depend on electron transfer through the coupling
site and have no relation to membrane energization.

Up to now we tacitly agreed that the carotenoid shift is caused by the effect of
electrochromism in a field of membrane potential. As a matter of fact, this notion
not only lacks proof but is rather doubtful. Electrochromism is based on a series
of phenomena connected with the perturbation effect of an external electric field
on the positions of electron levels, transition probabilities, and the orientational
distribution function of absorbing molecules. A discussion of the electrochromism
theory can be found, e.g., in [3.83]. For any reasonable electric field this perturbation is relatively small and can be accounted for within the realm of conventional first-order quantum-mechanical perturbation theory. The spectral changes induced by an external electric field are, therefore, in principle indistinguishable
from the changes induced by any small perturbation, e.g., conformational strains including those caused by the appearance of new local charges in the neighborhood of
the chromophore group. It is appropriate to remember here that there are no differences between electrical and mechanical forces at the molecular level [3.84].

From the aforesaid it follows that Δψ estimations based on the "carotenoid method"
are always overstated. It is even impossible to evaluate the resultant error: the
"true" (if one can call it "true", see Sect.3.3) transmembrane electrical potential
can have any value from that almost coinciding with the measured one to zero.

An ingenious new method of measuring electric potential differences generated
between water phases by the artifical and the natural membrane systems with incorporated energy-transducing protein complexes has been developed in SKULACHEV's
laboratory [3.85,86]. This method can, probably, provide qualitative information
concerning the generation of membrane potential.

The most direct method is, naturally, the immediate measurements of $\Delta\psi$ in sub-cellular organelles by means of reversible microelectrodes. Such measurements are performed only in few laboratories. Microelectrode measurements of $\Delta\psi$ values in mitochondrial systems have been performed for many years in the Tedeshi's laboratory in the USA. The precise technique developed by Tedeshi and his co-workers was de-scribed in detail in [3.87].

To be sure, it is rather dangerous to have to work with such small objects as mitochondria: they can easily be damaged by the inserted microelectrode. Tedeshi's opponents usually point out this very possibility [3.88]. Tedeshi noted, however, that the electrical resistance of mitochondrial membrane measured during these ex-periments is about 2 ohm.cm^2, i.e., it corresponds to the electrical resistance of other cell membranes [3.57,89]. The results obtained by the Tedeshi group are usually questioned because their $\Delta\psi$ values are, as a rule, by an order of magnitude lower than those obtained by the distribution of penetrating ions and, moreover, often have "wrong" signs ("+" inside, "-" outside). Results have, however, been published showing similar low $\Delta\psi$ values obtained by other methods [3.90,91].

Much fewer objections from the adherents of the chemiosmotic concept have been raised against the microelectrode measurements of $\Delta\psi$ values in photosynthetic sys-tems [3.92,93]. Although $\Delta\psi$ values presented in these works are as low as Tedeshi's values for mitochondria, and although the probability of mitochondrial membrane damage by a microelectrode is not higher than that of thylakoidal membrane damage, these results have been met with much less hostility, probably owing to the fact that in chloroplasts, in contrast to mitochondria, the main component in $\Delta\mu_{H^+}$ is, according to the orthodox chemiosmotic concept, not $\Delta\psi$ but ΔpH.

This short discussion concerning the existing methods of measuring transmembrane electric potentials in energy-transducing biological systems shows that these methods are, as a rule, inadequate to the task in question. The only method in which we measure what we want to measure, the microelectrode method, can be always called in question (which perhaps, is not always well grounded) due to the high probability of membrane damage.

3. 4. 2 ΔpH, $\Delta\psi$, $\Delta\mu_{H^+}$ Values and Phosphorylation in Membrane Energy-Transducing Systems

Bearing in mind all that has been said above about measurements in membrane struc-tures, let us now consider the results of quantitative estimations of the proton electrochemical gradient ($\Delta\mu_{H^+}$) and its components in phosphorylating mitochondria, chloroplasts, and chromatophores. When one reads the numerous papers concerning these measurements, it is difficult to avoid the impression of a permanent "energy crisis". Summing up the energy contributions of ΔpH and $\Delta\psi$ according to (2.5), the authors generally obtain $\Delta\mu_{H^+}$ values not large enough to ensure ATP synthesis. The energy shortage can be large or small, but it is inevitable. Relevant data can be found in numerous reviews [3.2,6,26,57]. We shall cite here several typical examples.

In MITCHELL's work [3.58], the maximum $\Delta\mu_{H^+}$ value obtained for mitochondria in the state of respiratory control (state 4 according to Chance's classification) was equal to 230 mV. It means that the system could spend on ATP synthesis 460 mV (according to the chemiosmotic concept the formation of one ATP molecule requires the passage of two protons through the channel of ATP synthetase). The authors took the free energy of ATP synthesis in the state 4 mitochondria to be equal to 540 mV, but they did not attached any great importance to this energy shortage and considered the agreement satisfactory.

When analyzing these data the following facts must be taken into account.

1) According to more reliable data [3.1] the phosphate potential of state 4 mitochondria is about 630 mV.

2) If the chemiosmotic concept is accepted we must take into account that phosphorylation proceeds not at equilibrium but at a steady state in a thermodynamically open system. A part of energy liberated due to μ_{H^+} decrease will, therefore, inevitably undergo dissipation.

3) The osmotic effects have not been taken into account. The matrix free water volume must increase because of K^+ ions being pumped into the matrix (Sect.3.4.1). This must lead to an overstated $\Delta\psi$ value determined according to (3.21).

The results obtained in [3.58] show that considerable contribution to the measured $\Delta\mu_{H^+}$ values is made by $\Delta\psi$: in one of the experiments $\Delta\psi$ determined by K^+ distribution in the presence of valinomycin was ~199 mV, and $\Delta\mu_{H^+}$ value was ~230 mV. In the preceding section it was shown that this method of $\Delta\psi$ estimation is rather doubtful and certainly gives strongly overstated $\Delta\psi$ values.

The data listed in [3.58] show that under all the conditions of experiment the amount of K^+ ions transferred into the matrix is quite close to the amount of protons transferred out of the matrix in the course of ETC functioning. In the case of a membrane with poor proton permeability this is indicative of the desicive role played by Donnan's mechanism of K^+ ions redistribution (Sect.3.4.1). This redistribution leads to the formation of an electric potential ("-" inside), and cannot be regarded as its consequence.

It can be calculated, using the data of [3.58] and formula (3.23), that Donnan's potentials which should be observed at the zero value of the initial "external" $\Delta\psi$ are not connected with Donnan's effect (similar calculations have been made by TEDESHI [3.66]). Subtracting from these Donnan's potentials the $\Delta\psi$ values obtained in [3.58] according to (3.21), we estimate "true" $\Delta\psi$ values existing in mitochondria under various experimental conditions.

The authors of [3.58] carried out their experiments with mitochondria in the presence of valinomycin whose amount was large enough to make all K^+ ions freely penetrating and predominant relative to the other penetrating ions present. Experiments of four types have been carried out with the following types of mitochondria.

1) Mitochondria have been initially improverished of K^+ ions in the external medium (low $[K^+]_o$ value) (prior to switching on the ETC). A special chelator was added to decrease the resultant ΔpH.

2) Mitochondria with initially low $[K^+]_o$ value.

3) Normal mitochondria.

4) Normal mitochondria with increased $[K^+]_o$ value.

The results of calculations can be seen in Table 3.1.

Table 3.1. Membrane-potential values for mitochondria under various conditions (recalculated using the data obtained in [3.58])

Type of experiment	$[K^+]_o$ [mol]	$\Delta\psi_1$ according to 3.58 [mV]	$\Delta\psi_2$ according to (3.22) [mV]	$\Delta\psi^a = \Delta\psi_1 - \Delta\psi_2$ [mV]
1	1.82×10^{-5}	199	194	+ 5
2	3.38×10^{-5}	171	184	- 13
3	7.25×10^{-4}	139	120	+ 19
4	9.62×10^{-3}	83	60	+ 23

[a]Positive $\Delta\psi$ values correspond to the directi on: "+" outside, "-" inside.

TEDESHI [3.66] in his analysis of data published in [3.58] obtained similar results. What has been said above leads us to the conclusion that $\Delta\mu_{H+}$ values obtained in [3.58] cannot ensure ATP synthesis in accordance with the chemiosmotic mechanism. The energy deficiency of chemiosmotic mechanism has also been noted by authors of later works [3.59,60].

A beautiful study, the results of which have been recently published [3.94a], probably gives a definitive answer to the question of whether $\Delta\mu_{H+}$ values measured in phosphorylating mitochondria are compatible with the requirements of the chemiosmotic concept. The measurements of all potentials, phosphorylation substrates and matrix-free volumes were made on intact cells of mice neuroblastome. Membrane potentials were determined with microelectrodes, as well as using the distribution of labeled penetrating ions taken in indicator quantities. The results of both methods practically coincided. Under the conditions of these experiments the synthesis of one mole of ATP required as a minimum 530 mV, whereas the maximum measured $\Delta\mu_{H+}$ value was 143 mV. This discrepancy cannot be explained by the orthodox chemiosmotic concept.[3]

Let us now turn to the experimental data obtained for photosynthetic systems. The study of the relationship between photophosphorylation in chloroplasts or chromatophores and the components of proton transmembrane electrochemical potential has been

---

3 Quite recently WILSON and FORMAN [3.94b] published data according to which a decrease of ΔpH values in mitochondria down to zero does not affect phosphorylation, electron transfer and the $\Delta\psi$ values.

for many years the favorite pastime of biophysicists and biochemists in many laboratories throughout the world. It is impossible even to mention all the works
concerning this problem.

The energy requirements of ATP synthesis in chloroplasts were established by
KRAYENHOF [3.95] who obtained $\Delta G \approx 15\text{-}17$ kcal/mol. Krayenhof measured the phosphate
potential

$$P = \frac{[ATP]}{[ADP][P_i]}$$

under conditions of saturating illumination and obtained $P \approx 30000$ mol^{-1}, which
corresponds to ~ 6.0 kcal/mol as the addition to the standard free energy value
ΔG^0 in (2.7). GIERSCH et al., however, have recently shown [3.96], that such P
values can be observed only when chloroplasts are able to exchange their adenylates
(ATP and ADP) freely with the environment. At physiological P_i concentrations (from
4 to 15 mmol [3.97,98]), P value of 30000 mol^{-1} corresponds to [ATP]/[ADP] ratio
from 120 to 450. Careful determinations of this ratio in intact chloroplasts give,
however, much lower values: 2-4 in the light and ~ 1 in the dark [3.99-101]. This
is caused by intensive ATP consumption in the course of the reduction of substrates
and CO_2, so that even with the highest light intensities and, consequently, at the
fastest electron transport, photophosphorylation is incapable of maintaining high
steady-state ATP concentrations [3.96]. Large values of phosphate potential can be
reached by either adding the surplus of a substrate not requiring ATP for its reduction (e.g., nitrites, oxalacetate, phosphoglycerate) or by working with partly
wrecked chloroplasts with a damaged envelope but intact thylakoidal membrane.
Kragenhof and many other authors evidently worked with such chloroplasts. As a
matter of fact, it does not call in question the comparisons of the $\Delta\mu_{H^+}$ values (if
measured correctly) with the phosphate potential values in these works: phosphorylation in chloroplasts can proceed even at much higher P values.

The following important results obtained in [3.96] must be emphasized. Phosphorylation rate in intact chloroplasts does not depend on P over a wide range of
its values. Chloroplasts are able to synthesize ATP even at the highest P values
reached: $P > 8 \times 10^4$ mol^{-1}. Considerable changes in $\Delta\mu_{H^+}$ values do not influence
either measured P values or, consequently, the phosphorylation process. There are
no equilibrium relations and no connections through mass action law between the processes determined by $\Delta\mu_{H^+}$ and P. The authors conclude that in intact chloroplasts
$\Delta\mu_{H^+}$ is not high enough to ensure ATP synthesis, even if we assume that ATP is formed
by utilizing the energy liberated during the passage of three (and under certain
conditions four) protons through the ATP-synthetase channel.

Paper [3.96] contains many important results concerning the mechanisms regulating
electron transport and phosphorylation in photosynthetic systems. Their consideration goes beyond to scope of the problems discussed in this book. It is, however,
appropriate to note here that in intact chloroplasts the [ATP]/[ADP] ratio does not

determine the electron transfer rate. The authors suggested that electron transport regulation is ensured by the intrathylakoidal pH values. It has been shown to be true in work carried out recently in our laboratory (see below).

Two important notes made by the authors of [3.96] at the end of their paper must be underlined here. They point out with good reason that in fact the relationship between $\Delta\mu_{H^+}$ and phosphorylation must be discussed in terms of nonequilibrium thermodynamics, and that the discrepancy will, therefore, be much more striking.

The problem of energy dissipation during the coupling of chemical reactions in the steady states of open systems has been considered in detail in a very thorough study by STUCKI [3.102]. Although the analysis was carried out for oxidative phosphorylation in mitochondria, the results obtained are valid for any energy-transducing process if it is treated in the realm of nonequilibrium thermodynamics in a linear approximation.

The principal conclusion is that in actively phosphorylating systems the optimal course of phosphorylation process (the most complete possible utilization of energy liberated during the energy-donating reaction and the largest possible yield of the energy-accepting reaction) inevitably involves considerable (not less than 50%) dissipation of energy liberated in the course of the energy-donating reaction. The $\Delta\mu_{H^+}$ values under the conditions of continuous proton and adenylate flows in a steady state must at least be twice as high as the P values. Other authors [3.103] arrived at the same conclusions.

The second note concerns the results of ENSER and HEBER [3.104], according to which the acidity of the intrathylakoidal space of dark-adapted chloroplasts is by about one pH unit higher than that of stroma. This ΔpH is insensitive to uncouplers and cannot, therefore, participate in ATP formation. The discrepancy between $\Delta\mu_{H^+}$ and P values must thus be even greater than that noted in [3.96].

In a study of the electron transfer regulation between two photosystems in chloroplasts of green plants, TIKHONOV et al. [3.105] have been able to show that in intact chloroplasts and whole leaves the rate of electron transfer between plastoquinone and the active center of photosystem 1, $P700^+$, in the 5-9 pH range is determined solely by the pH value within the intrathylakoidal space. The transfer rate can be conveniently measured by the EPR method (I cannot describe here the technical details) allowing the intrathylakoidal pH value to be estimated without any additions to the chloroplasts or whole leaves. The ΔpH dependence on the switching on of phosphorylation has been studied at different pH values. It was found that phosphorylation leads to a large drop in ΔpH value. Under optimal photophosphorylation conditions, at pH 7.5, when the ATP synthesis rate at a steady state under white light illumination is maximal, ΔpH value does not exceed 0.5. This means that in chloroplasts the transmembrane difference between proton concentrations is not necessary for ATP synthesis.

When the chemiosmotic concept, and experiments which do not conform to it, are discussed in the presence of its adherents, standard easily predictable situations always arise. If an experiment shows that phosphorylation can proceed without ΔpH, then the statement that follows is almost inevitably: "so the large $\Delta\psi$ value is responsible". It is rather convenient that (3.5) contains not one but two terms. There are, however, only two, and it is not difficult to check both.

The transmembrane electric potential between intrathylakoidal space and stroma has been measured using different techniques. Direct measurements with microelectrodes [3.92,93] in a steady state give negligibly low $\Delta\psi$ values (as a rule < 10 mV). Similar results were obtained with the penetrating ions in indicator quantities [3.106]. WREDENBERG [3.93] comes to the conclusion that evidently the main energy source in chloroplasts is ΔpH and not $\Delta\psi$. Higher $\Delta\psi$ values are, as a rule, obtained with the electrochromism method using endogeneous carotenoids or specially added dyes. It has already been stressed that this method gives rise to serious doubt. It should be noted that $\Delta\psi$ values determined by this method exhibit a tendency to decrease with years. Thus, the initial $\Delta\psi$ values, estimated by carotenoid shift were ~200 mV but already in 1974 these values decreased by half (all data and references can be found in [3.23]). Measuring the spectral shift of added oxanol dye, SCHURMANS et al. have estimated the steady-state $\Delta\psi$ value as not exceeding 50 mV [3.107]. The disagreement between the results obtained by electrochromism and by other methods is now, as a rule, explained by the fact that electrochromic shift feels a "surface potential", while the other methods feel the potential difference between bulk phases [3.23]. As already mentioned, the electrochromism method cannot differentiate between the volume or surface charges and any local charge or a sterical constraint. It is clear, therefore, that even accepting $\Delta\psi$ values presently determined with the help of electrochromism method as true ones, one cannot obtain $\Delta\mu_{H^+}$ values large enough to satisfy the requirements of chemiosmotic concept.

Photophysical and photochemical methods allow us to study the kinetics of electron transport and photophosphorylation in a broad time interval. Especially interesting opportunities are provided by the use of saturating light flashes of such a short duration that only one electron is transferred from water to $NADP^+$ in response to one flash. It is well known that the active center of photosystem 2 has a peculiar latent period: after its excitation and the primary act of charge separation the center remains unable to realize the next act during ~0.6 ms [3.108]. This makes it possible to study the ATP synthesis in response to the passage of electrons through chloroplast ETC one by one (similar conditions can be realized for the chromatophores of photosynthetic bacteria).

Illumination of chloroplasts with short single flashes practically does not lead to any ΔpH formation (pH changes do not exceed 0.01-0.03 pH units, the rise time is ~20 ms, and the decay time, ~1 s [3.109,110]). The formation of transmembrane electric potential after illumination of chloroplasts and chromatophores with short saturating flashes has been studied by many scientists. Most of them applied the electrochromism method (strictly speaking, they measured the intensity of the shifted carotenoid absorption band at ~520 mn). It was shown in the preceding section that this method can at best give only the upper limit of $\Delta\psi$ value. A survey of corresponding data can be found in [3.23]. Analysis of these data, on overwhelming part of which had been obtained in his laboratory, led Witt to the conclusion that $\Delta\psi$ measured with "carotenoid shift" in chloroplasts in response to a single saturating flash of microsecond duration cannot exceed 50 mV. Taking into account the practical absence of ΔpH under these conditions, we can conclude that in the realm of the chemiosmotic concept, phosphorylation in response to a single saturating flash of microsecond duration is impossible.

In order to verify this conclusion, a systematic study was carried out in our laboratory in 1976-1977 [3.6,111-113]. Similar data (mainly on chromatophores) were independently obtained in the works of other groups [3.114-117]. In our experiments, chloroplasts were illuminated by short white light flashes of 10 μs duration and of saturating intensity (control experiments with laser flashes of 30 ns duration have given the same results). The sensitivity of the method used (inclusion of radioactive P_i labeled with phosphorus into ATP) made it possible to measure phosphorylation in response to a single flash, as well as after a series of single flashes. The following principal results have been obtained.

1) In the case of dark adapted chloroplasts, the transfer of one electron has led to the synthesis of 0.4-0.5 ATP molecules per ETC. A partial inhibition of electron transfer with DCMU addition in various concentrations or by diminishing the flash intensity has led to a reduced ATP yield in response to a single flash in proportion to the decrease in the number of electrons transferred. This means that phosphorylation in chloroplasts does not require a cooperative action of coupling sites localized in different ETC. No threshold intensity could be found.

2) After the first phosphorylation act, subsequent flashes gave a higher ATP yield per flash, which after several phosphorylation acts reached the maximum value of ~1 ATP per ETC per flash. It is the phosphorylation process, and not the electron transfer, which is important: photoinduced electron transfer in the absence of phosphorylation substrates does not lead to an increase of the ATP yield per flash after phosphorylation substrates have been subsequently introduced into the system. The role played by phosphorylation substrates in these phenomena cannot, however, be reduced to the appearance of free ATP molecules, because ATP addition does not in itself lead to the increase of photophosphorylation yield.

The effect of the fast ATP yield increase on subsequent flashes can be observed only if the dark gap ($\mathscr{T}_d$) between the flashes is sufficiently short (at room temperature $\mathscr{T}_d \approx 1.0$ s). At longer $\mathscr{T}_d$ the system "forgets" the preceding phosphorylation acts, and at $\mathscr{T}_d > 6$ s the ATP yield per flash is practically equal to that for dark-adapted chloroplasts.

3) The dark lifetime of the state capable of enhanced phosphorylation, as compared with the dark-adapted chloroplasts depends but slightly on temperatures from 10° to $40^\circ C$. At the same time, the initial distribution of centers between two types (capable of immediate phosphorylation and requiring several sufficiently fast preliminary phosphorylation acts in the centers of the first type) changes significantly with temperature, shifting in favor of the second type of centers as the temperature increases.

4) In the presence of silicomolybdate serving as electron acceptor and of DCMU used to inhibit electron transport to photosystem 1, i.e., when only photosystem 2 is functioning, the maximum ATP yield and the ATP yield after the first flash decrease proportionally, but the ratio between these yields remains the same as in the case of normal chloroplast functioning. Qualitatively the same result has been observed in the case of photophosphorylation by photosystem 1 only with an exogeneous electron donor. It means that photophosphorylation in chloroplasts does not require cooperative action of the coupling sites localized within one ETC.

5) The ATP yield per electron transferred does not depend on the [ADP]/[ATP] ratio within the interval from 10^{-2} to 10^2.

It must be emphasized that the experiments in all cases were performed in such a way that only new synthesized ATP has been determined, but not ATP initially bound in chloroplasts. Neither can the results obtained be explained by adenylatkinase activity. It was possible to synthesize hundreds of ATP molecules per ETC with one and the same sample, repeating single flashes of microsecond duration with dark gaps of any duration from 0.3 s to more than 10 s.

The most important conclusion that can be drawn based on the above results is that the coupling of "act per act" type between enervy-donating and energy-accepting reactions in the course of photophosphorylation in chloroplasts (and evidently, in chromatophores of photosynthetic bacteria) is realized straight away, long before the system reaches a steady state. The analysis performed in Sect.3.2 implies that this fact rules out not only chemiosmotic mechanism in its orthodox formulation, but also any coupling mechanism based on the mass action law. Such coupling requires a molecular machine of enthalpy type, with an excitation of specific degrees of freedom which interact but slightly with thermal ones.

Analysis of these results allows us to formulate the following preliminary conclusions. In the dark-adapted chloroplasts a certain fraction (at room temperature about one fourth) of all the coupling sites exists in such a state that the passage of one electron through the site liberates energy sufficient to ensure the synthesis of one ATP molecule. The energy liberated during the transfer of one electron through other coupling sites is not high enough to ensure ATP synthesis. After several phosphorylation acts the thylakoid membrane is transformed into a new state in which the number of phosphorylating coupling sites increases to a maximum value corresponding to half of all the existing sites (these estimations are based on the widely accepted point of view that a chain of noncyclic electron transport contains two coupling sites). This "energized" state of the membrane lives for ~1 s at room temperature.

The widely accepted view, according to which the synthesis of one ATP molecule requires the transfer of two electrons through a coupling site, thus seems to be wrong (at least for pulse illumination regime). Simply in "energized" thylakoid membranes only about one half of the coupling sites exist statistically in a "phosphorylating state". In relaxed, dark-adapted chloroplasts this fraction decreases to about one fourth. This implies, of course, that equilibrium values of redox potential differences between the ETC carriers bear little relation to their functioning as energy donors.

It should be noted here that in the realm of classical equilibrium thermodynamics, which in the final analysis is the basis of the chemiosmotic concept, it is impossible to obtain the necessary $\Delta\mu_{H^+}$ values in the energy-transducing membranes of mitochondria and chloroplasts owing to ETC electron transfer. Indeed, the proton pumps, that transfer hydrogen ions into thylakoidal inner space or out of mitochondrial matrix, use energy liberated by electron transfer between certain pairs of the neighboring ETC carriers. Therefore, even with a 100% efficiency of the coupling process, proton transfer must cease as soon as the value becomes equal to the redox potential difference of the particular pair of carriers, which ensures the functioning of this particular coupling site (it must be borne in mind that coupling sites function independently). We have seen in Sect.2.4 that there is no pair of neighboring carriers in chloroplasts as well as in mitochondria with a high enough redox potential difference. Direct experiments carried out recently have, indeed, shown that, at least in the ETC of certain bacteria, ATP synthetase functions irreversibly [3.118].

According to very interesting series of papers [3.119-121] even the first initial postulate of the chemiosmotic concept concerning the necessity of closed vesicles may be called in question. The authors [3.119-121] have shown that submitochondrial particles after certain mild treatment lose their vesicular structure but preserve their phosphorylation activity.

3. 4. 3 Membrane Phosphorylation Caused by Various Types of "Strokes"

Excellent experiments carried out by JAGENDORF and URIBE and published in 1966 [3.122] have been for many years considered as one of the most important and direct proofs of the chemiosmotic concept. In these experiments chloroplasts were incubated in the dark at pH 4 long enough (20-30 s) to reach this pH value on both sides of the thylakoid membrane. After that, an alkaline solution containing ADP and P_i was rapidly added so that the outer pH value almost immediately became ~8. ATP synthesis proceeds during 2-3 seconds after this "alkaline stroke", and up to 100 ATP molecules per one ETC have time to be formed. The interpretation of these experiments seemed to be selfevident. After a fast addition of alkaline solution a proton concentration difference ΔpH appeared between the inner and the outer thylakoidal spaces. The sign of pH gradient corresponds to the requirements of the chemiosmotic concept (increased intrathylakoidal proton concentration). Energy liberated during the passage of proton surplus through the ATP synthetase channel ensures ATP synthesis.

Data have repeatedly appeared in literature calling in question this straightforward interpretation of Jagendorf's and Uribe's experiments. It was, e.g., noted that the rate of ATP formation does not depend in this case on ΔpH but is determined by the absolute pH values of the inner and outer thylakoidal spaces [3.123]. Such small discrepancies between experiments and theory do not, as a rule, embarrass the adherents of the latter. In this connection the work of GOULD et al. [3.124] is of great interest and deserves a through consideration. In this beautiful experimental study phosphorylation was conducted in an artificial system; ATP-synthetase complex from spinach chloroplasts was incorporated into the envelope of phospholipid membrane vesicles. Under standard experimental conditions there was a solution with a low buffer capacity (at pH 8) inside the vesicles and a solution with a high buffer capacity (at pH 8) outside the vesicles. The solutions contained phosphorylation substrates (ADP and P_i). The authors have set themself the task to create ΔpH (i.e., the motive force of phosphorylation according to the chemiosmotic concept) without any additions. In this case ΔpH appears as a result of pulse γ-radiolysis of the samples, known to produce protons in a water medium. The doses used (500 pulses in 4 seconds) led the inner pH to decrease by ~1.2 units, while the outer pH value, because of the much higher buffer capacity, did not change. Two methods of vesicle preparation were tried. According to the first method, vesicles are prepared with the help of an emulgator, Na-cholate. The above procedure led, in this case, to the formation of 8 ATP molecules per one enzyme complex. According to the second method, vesicles are prepared by the acoustic action, and the above procedure led to the formation of 2 ATP molecules per one enzyme complex. The paper gives quantitative data only for vesicles prepared by the second method. Various control experiments (without vesicle formation, without enzyme, without ADP) gave, just as it was to be expected, negligible ATP yields, about one order of magnitude smaller than the usual yield of ~2 ATP molecules per enzyme complex. The authors have, however, carried out one more control experiment: the system contained one and the same high-capacity buffer within as well as outside the vesicles (the same solution that was only outside during the main experiments). According to the authors there is no pH change under these conditions, and naturally, no ΔpH formation can be realized. The ATP yield, however, was in this case of the same order of magnitude as that obtained in the main experiments (the yield decreased by less than half). It is quite clear that the results described in this paper show that ΔpH is *not* the driving force of phosphorylation in this case. The authors, however, draw an opposite conclusion completely neglecting the result of their principal control experiment.

MAGNUSSON and McCARTY [3.125] found that ADP bound within chloroplasts can form ATP with a 25% yield after an acid stroke. This effect was reproduced and carefully studied in our laboratory [3.126]. If the stroke with a strong acid was fast enough to change the pH value from 7.4 to ~1.6 in less than 1 s, then, in the presence of ADP and P_i, the yield of ATP was exactly equal to the maximum yield per one saturating flash in energized chloroplasts in the previously described experiments [3.111-113] (special experiments with one and the same sample, first illuminated by saturating flashes and then exposed to the acid stroke). Careful control experiments completely exclude the possibility of explaining these results by the liberation of previously formed tightly bound ATP or by the adenilatekinase activity. This procedure cannot, naturally, be repeated with one and the same sample: under the action of acid chloroplasts become completely inactivated. Evidently, phosphorylation has enough time to be completed during the denaturational inactivation of chloroplasts. This result cannot be explained by ΔpH formation: even if ΔpH is formed, its sign is opposite to the requirements of the chemiosmotic concept.

Chloroplast experiments of Jagendorf and Uribe have been repeatedly reproduced in many laboratories throughout the world. It would be quite natural to try to reproduce this experiment with mitochondria. As the ΔpH and Δψ vectors in mitochondria have the signs opposite to those in chloroplasts, realization of ATP synthesis in mitochondria under conditions analogous (according to the chemiosmotic concept) to those in Jagendorf's and Uribe's experiments, requires not an acid-base but a base-acid transition. Judging by [3.127] such attempts were made in various laboratories right away after the results of Jagendorf and Uribe became known, but were unsuccessful. As far as I know only one paper has been published concerning pH-induced ATP formation in mitochondria [3.127]. The experiment design realized in Mitchell's laboratory was as follows. Rat liver mitochondria were incubated at pH 7.0 in the presence of ADP and P_i. The pH was abruptly increased to 9.0, and after some time decreased again to 4.4. ATP content in the sample as a result of this treatment increased at maximum by ~500 μmol per 1 mg of mitochondrial protein (i.e., by approximately 3 ATP molecules per ETC). The preliminary pH increase up to 9.0 was, in the authors' opinion, necessary to ensure a large enough ΔpH value after the subsequent pH decrease: smaller ΔpH values were ineffective, and at pH values lower than 4.4 mitochondria become inactivated.

The possibility of ATP synthesis in mitochondria due to fast enough pH change was minutely investigated in our laboratory (these experiments were partly described in [3.128]). The results stated below were obtained with rat liver mitochondria having after isolation respiratory control coefficients from 4 to 6. Experiments carried out with beef heart mitochondria gave the same results. ATP was determined radiometrically using a method based on the reaction between ATP and (^{14}C) glucose in the presence of hexokinase and the subsequent isolation of (^{14}C) glucose-6-phosphate. All the necessary precautionary measures were taken and control experiments were carried out in order to exclude any hampering factors, such as adenilatekinase activity of the sample or the possibility of the effect being explained by liberation of the previously formed ATP molecules.

Two types of experiments have been carried out. In the experiments of the first type we used actively phosphorylating mitochondria. A buffer solution containing phosphorylation substrates was added to anaerobic mitochondria suspension so as to increase fast enough ($\leq$ 0.5 s) the pH value of the suspension from ~7.5 to ~8.5. The initial suspension contained succinate and the added buffer, oxygen. Introduction of oxygen into the anaerobic suspension leads to the initiation of respiration, and in the coupled mitochondria to normal ATP synthesis, which is recorded together with the ATP synthesis caused by pH jump. This normal ATP synthesis could be accounted for by carrying out the same experiment at initial or final pH values of both the suspension and the buffer solution without pH jump. In this way it was shown that a fast pH increase by one unit, when the sign of the pH gradient to be created is reversed relative to the requirements of chemiosmotic hypothesis, leads to additional ATP synthesis by the coupling mitochondria. A fast decrease of pH value by 1-3 units (i.e., creating the pH gradient of the "proper" sign) did not lead to any additional ATP synthesis.

In the experiments of the second type the phosphorylating activity of mitochondria was completely suppressed either by "aging" or by a freezing and thawing procedure [3.129].[4] The respiratory activity of mitochondria after this treatment is preserved. The results did not depend on the method of inactivation. ATP synthesis due to fast pH increase by uncoupled mitochondria unable to ensure normal oxidative phosphorylation was studied with special thoroughness. It was found that a sufficiently fast pH increase ($\mathcal{T} \leq$ 0.5 s) leads in this case to the synthesis of 2-3 ATP molecules per ETC. The maximum ATP yield is reached if the amplitude of pH jump exceeds 0.7 units. ATP yield does not depend either on the initial or the final pH value provided the system passes pH values of 8.1-8.3 in the course of pH change. For instance, ATP yield is the same with a pH increase from 6 to 10 as with an increase from 7.8 to 8.5. A pH decrease never leads to ATP formation.

Various control experiments (use of inhibitors, different respiration substrates) have shown that a necessary condition for ATP synthesis with pH jump is the intermediate redox state of ETC components (perhaps including ATP synthetase). Neither in the state of complete oxidation nor in the state of complete reduction can mitochondria suspension synthesize ATP after a jump-like increase in pH. It is possible to obtain in this way scores of ATP molecules per ETC with one and same sample by changing pH repeatedly from, e.g., 7.5 to 8.5 and back again.

---

4 See, also, the detailed description of these experiments in [3.131].

Oligomycin completely inhibits this pH-jump induced ATP synthesis. It means that we observe in this case the ATP formation catalized by a mitochondrial ATP-synthetase.

It is clear thus that in mitochondria, as well as in chloroplast, ATP synthesis can be caused by a jump-like increase (but not by a decrease) of the pH value. We can, therefore, conclude that the coincidence of the sign of pH gradient with the requirements of chemiosmotic concept in [3.122] cannot be regarded as a proof of the latter.

An obvious interpretation of these results can be briefly formulated as follows. A jump-like increase in the pH value leads to instantaneous ionization of certain acid groups of ETC proteins (and/or ATP synthetase) with pH from 8.1 to 8.3. Sufficiently fast ionization essentially represents the formation of a conformtionally out-of-equilibrium state of protein (Chap.4) in the course of whose relaxation the condensation of ADP with P_i takes place. The immediate energy source in this case is thus an acid-base reaction of the type

$$AH + OH^- \rightarrow A^- + H_2O \quad , \tag{3.24}$$

where AH stands for acid groups with pH values between 8.1 and 8.3. Relaxation processes of this type and physical aspects of their participation in bioenergetic phenomena will be considered in Chaps.4 and 5).

An ATP synthesis in membrane systems can also be observed after "electric strokes", i.e., due to an abrupt rise of electric field across the energy-transducing membrane. The experiments have been carried out on the thylakoids [3.130], mitochondria and submitochondrial particles [3.131], and on the liposomes modified by H^+-ATPase [3.132]. The most detailed study carried out recently by VINCLER and KORENSTEIN [3.133] on chloroplasts has shown that a rectangular electric pulse of 0.3-3.0 ms duration and with a field amplitude of ~1 kV leads to the synthesis of ~6 ATP molecules per ATP-synthetase. ATP formation proceeds for a period of seconds after the electric pulse and is not inhibited by gramicidine and other ionophores. It has been concluded that this electric-field-induced ATP synthesis cannot be explained by the creation of any transmembrane ion concentration gradient.

We have already seen that ATP synthesis can be ensured by "strokes" of other kinds, e.g., the stroke leading to chloroplasts acid denaturation [3.126], or the stroke by a pulse of ionizing radiation [3.124] (the "stroke" in this case is, probably, the sufficiently fast reduction of the metal-containing active centers of ETC and ATP synthetase, see Chap.4). There are data in literature of ATP synthesis in chloroplasts caused by the "acetone stroke" [3.134] and in mitochondria by the "temperature stroke" [3.135]. One can get the impression that ATP synthetase is a punching machine able to perform only one operation: to synthetize ATP from ADP and P_i in response to almost any perturbation, whatever strokes are used. I think that experimental data and theoretical consideration presented in this chapter compels us to treat with certain doubt the chemiosmotic concept of membrane phosphorylation so popular nowadays.

4. Proteins as Molecular Machines

The analysis of various biological energy-transducing systems and processes carried out in preceding chapters shows that in all cases we are dealing with the functioning of certain mechanical constructions, machines of rather small dimensions down to molecular ones, which ensure the elementary acts of coupling between energy-donating and energy-accepting chemical reactions. Expressions such as "a protein is a machine", "an enzyme is a machine", and "energy-transducing constructions" have now become stock phrases. Usually, however, these statements remain just words, and do not bear any definite meaning. The main reason for that is the fact that in spite of all the talk concerning the "machineness" of proteins in general, and of enzymes in particular, most scientists, when considering chemical properties of biopolymers, use standard approaches of chemical thermodynamics and chemical kinetics developed over the last 100 years for reactions of low-molecular compounds in gaseous phase and dilute solutions. These approaches are based essentially on classical statistical physics of ergodic systems, i.e., on the assumption that these systems have only thermal degrees of freedom with fast enough energy exchange between them. If, however, biologically active macromolecules are machines, it means that during their functioning specific mechanical degrees of freedem interacting only weakly with thermal ones must be excited. Only the lack of such specific degrees of freedom can justify the use of Van't Hoff equation

$$K = \exp\left(\frac{\Delta S}{R}\right) \exp\left(- \frac{\Delta H}{RT}\right) \tag{4.1}$$

when considering the thermodynamic characteristics of biological systems, and an Arrhenius approach leading in the realm of the activated complex theory to equation

$$\tilde{k} = x \, \frac{kT}{h} \, \exp\left(\frac{S_a}{R}\right) \exp\left(- \frac{E_a}{RT}\right) \tag{4.2}$$

when considering the kinetic characteristics of biological systems. It is easy to show [4.1,2] that both these equations are valid only if the following well-known thermodynamic expressions hold true:

$$\frac{\partial \Delta H}{\partial T} = R \, \frac{\partial \Delta S}{\partial T} \tag{4.3}$$

$$\frac{\partial E_a}{\partial T} = R \frac{\partial S_a}{\partial T} \quad . \tag{4.4}$$

Formulas (4.3,4) have a quite clear and simple physical meaning. Temperature changes of the entropy and enthalpy values of the initial and final reaction products, as well as those of activated complex (i.e., of a transient state at the top of the potential barrier) are wholly determined by the temperature changes of the Boltzmann distribution of molecules and activated complexes between their vibrational and rotational energy levels. In other words, it is assumed that temperature can change the spatial configuration of a particle either through the anharmonicity of molecular vibrations or through centrifugal distortions, but cannot change the molecular constructions in the above three states. Validity of this assumption in the case of complex biopolymer molecules is rather doubtful. Analysis of numerous experimental data [4.2] shows that the structure of a protein changes with temperature not in accordance with (4.3,4), but rather as a construction made of wires with different coefficients of thermal expansion for different wires.

What are we measuring in this case if we use equations of the Van't Hoff or the Arrhenius types? A detailed analysis can be found in [4.2]. Let us assume that the activation barrier changes with temperature "nonthermodynamically", i.e., without any relevant entropy change according to (4.4). In a narrow temperature range (all biochemical experiments for obvious reasons are carried out within narrow temperature ranges) any temperature dependence may be approximated by a linear one

$$E_a = E_a^0 + \alpha T \quad . \tag{4.5}$$

Introducing (4.5) into an equation of the Arrhenius type, e.g., into (4.2), we obtain

$$\tilde{k} = \frac{kT}{h} \exp\left(\frac{S_a - \alpha}{R}\right) \exp\left(-\frac{E_a^0}{RT}\right) \quad . \tag{4.6}$$

In this case, the experimental data will thus formally satisfy the Arrhenius equation, but we shall measure not the true activation parameters S_a and E_a, but effective quantities $S_a - \alpha$ and E_a^0. From (4.5) it is evident that the measured value E_a^0 is obtained by extrapolating a weak temperature dependence of E_a to 0 K, i.e., under the conditions of a biochemical experiment, by 300 K. This, naturally, may lead to physically meaningless values of activation parameters, which have nothing in common with their true values. The analysis of thermodynamic parameters ΔS and ΔH leads to similar conclusions.

There is a second tacitly accepted assumption closely connected with the first one. According to this assumption, the elementary act of a chemical reaction, chemical transformation of an individual molecule, takes place instaneously. The rate of a chemical reaction is determined not by the duration of the elementary act, but by the number of instaneous acts per second. Vibrational relaxation in a condensed phase takes only 10^{-12} - 10^{-13} s, and the molecules of the product acquire their equilibrium configurations immediately after the elementary act. This assumption is thus reduced to the postulate that in the course of a chemical reaction there are no molecules in the intermediate nonequilibrium configurations. Both these assumptions, as a rule, do not hold true for the reactions of complex protein molecules (as well as of other biologically active macromolecules). We shall begin our analysis of molecular machines in biological systems with the physical pecularities associated with their small dimensions.

4.1 The Physics of a Small Machine

In all textbooks on physics the following statement can be found: the optimum regime of the functioning of any energy-transducing device is a reversible process, i.e., a process so slow that at any moment of time the system is practically in the state of local equilibrium. It is customary to consider as an example a cylinder with a piston, in which the useful work (lifting a weight) is performed owing to gas expansion (Sect.3.2). The optimum regime of maximal efficiency is realized when the weight changes so slowly that at any moment of time the external pressure (weight per unit area) is equal to the internal equilibrium gas pressure. This condition of the optimum course of a process is, as a rule, automatically extended to cover biochemical processes. The proximity between the free energy changes of energy-donating and energy-accepting reactions in oxidative phosphorylation (i.e., the proximity between the phosphate potential and the overall ETC redox potential difference) is, for instance, considered as proof of thermodynamic reversibility of the oxidative phosphorylation process [4.3]. As far as I know, the repudiation of thermodynamic reversibility as an obligatory condition for enzymatic and bioenergetic processes was formulated in 1972 [4.4,5]. A detailed physical analysis of this problem was first carried out by GRAY [4.6]. The contents of this excellent paper will be briefly given here.

Thermodynamic reversibility of a machine process cannot be realized without a certain control, without determining the system's instantaneous state. In the above example with the weight being lifted as a result of gas expansion, the choice of a "proper" weight (i.e., such a weight that the external pressure is close enough to the gas pressure) requires the measurement of the piston positions, in other words, requires that either an operator or an automatic device obtains information about the state of a system at a given moment of time. Information, however, cannot be obtained "gratis" [4.7]. In order to determine coordinate x with precision $\pm\Delta x$ ($\Delta x > 0$), energy expenditure of

$$\Delta E \geq \frac{hc}{4\Delta x} \qquad\qquad\qquad (4.7)$$

is necessary.

Let us consider the functioning of a certain cyclically working machine able to perform mechanical work (as already noted, there is no difference between mechanical and electrical forces at molecular level [4.8]). Let the amount of energy entering the machine during one cycle, without accounting for the "information energy expenditure" (4.7), be equal to Q, and useful work executed during one cycle be equal to W. The resultant factor of energy conversion will be $\theta = W/(Q + \Delta E)$. Let us introduce the concept of the relative control value: $z = \Delta E/Q$. Within the classic approach $\Delta E \to 0$, $z \to 0$, and $\theta = W/Q$ is maximal when $W = W_{rev} = -\Delta G$ (change in the system free energy), i.e., when the process is thermodynamically reversible. For any real process $\Delta E > 0$, and factor θ is always less than its limiting value $\theta = W_{rev}/Q$. An increase in the precision of the "piston" position measurement, i.e., a decrease in Δx, leads, on the one hand, to an increase in the information energy expenditure (ΔE) (i.e., to a decrease in θ), but, on the other hand, will increase W, shifting its value to that of W_{rev}. There must thus exist certain "optimal precision", i.e., a certain optimum relative control value z_{opt} that gives the maximum factor of energy conversion Q. Elementary calculations carried out by GRAY [4.6] have shown that for macroscopic machines z_{opt} is extremely small, and the energy cost of a control accounts for a negligible fraction of the energy transduced by the machine in the course of one cycle. For example, if the process consists in the expansion of one mole of a gas, $Q = 4.10^4 J$, $z_{opt} = 10^{-14}$, and the information energy expenditure can be neglected in all estimations of thermodynamic characteristics. Thermodynamic reversibility is, thus, indeed seen to be an obligatory condition for the optimum functioning of macroscopic energy-transducing machines. Quite a different situation arises in the case of machines of molecular dimensions. During the functioning of one myosin bridge (Sect.2.1), the hydrolysis of one ATP molecule produces $Q \approx 7 \times 10^{-20} J$. If we require a rather modest degree of reversibility with W/Q = 0.9, then the calculations give $z = \Delta E/Q = 25$. It means that with $\sim 7 \times 10^{-20} J$ of energy entering our system during one cycle we must spend $\sim 2 \times 10^{-18} J$ per cycle to ensure this degree of thermodynamic reversibility. It is quite clear that for such small machines any form of control, which may allow us to conduct the process reversibly, is absolutely out of the question.

For molecular machines, therefore, the reversibility of a process does not ensure its optimal realization. GRAY wrote [4.6]: "rate is uncontrolled and determined by the mechanics (quantum or semiclassical) of the system itself." Machines of molecular dimensions cannot in principle function reversibly. If the energy coupling of chemical reactions, which is the basis of all bioenergetic processes, is indeed ensured by molecular machines, then the paths of the process in the forward and the backward directions cannot coincide.

A few words, now, on the irreversibility of processes taking place within one isolated system of molecular dimensions [4.9]. There is a widespread belief that quantum processes in an isolated molecule are obligatorily reversible. Indeed, according to the principle of microscopic reversibility the probabilities of forward and backward energy transfer between two states are always equal. There exist, however, situations when a complex system has many states some of which are indistinguishable during its functioning. The process in question can then consist in the transfer from one definite state (specially "prepared") to any of the indistinguishable states. To be sure, in the case of an isolated system whose overall energy is constant, the system will some time return to the initial state. For a sufficiently complex system, however, this "some time" may be not so soon, and the reversion of the system's state may never take place during the lifetime of a system or during the time occupied by our study. We shall see later that it is true for all bioenergetic processes. The aforesaid implies that in molecular machines that ensure energy-transducing processes in biological systems, i.e., in protein macromolecules and their complexes, essentially nonequilibrium states must appear. In the next section experimental evidence of the existence of such states will be described.

4.2 Conformationally Nonequilibrium States of Proteins

A molecule of ammonia NH_3 has the structure of trigonal pyramid with the nitrogen atom in the apex (Fig.4.1). What is the meaning of this statement? As every molecule, ammonia does not possess a fixed spatial structure. The valence and deformation vibrations of atoms relative to their equilibrium positions take place at any temperature. The frequencies of these vibrations are 10^{12} - 10^{13} s^{-1}, and approximately 1000 times more seldom the molecule is turned inside out, a quantum inversion transition occurs (Fig.4.2). One can thus speak not only of the spatial but also the dynamic structure of small molecules. The statement: "An ammonia molecule can exist in two equivalent states: a pyramid with the apex up and a pyramid with the apex down" has a physical meaning, because during the lifetime of a molecule in one of these states many atomic vibrations relative to their local (for this state) equilibrium positions have enough time to be completed. In the realm of quantum mechanics these states are, naturally, not stationary. A stationary state can be described as a superposition of nonstationary states (for a detailed analysis of the stationary and nonstationary systems of this type see [4.10]). The transition between two nonstationary states has, in this case, the nature of a quantum jump. To speak of "intermediate states", corresponding, e.g., to the arrangement of all four atoms of NH_3 in one and the same plane, is as meaningless as to speak of the transition trajectory of an electron between two stationary orbitals during the emission or absorption of a radiation quantum.

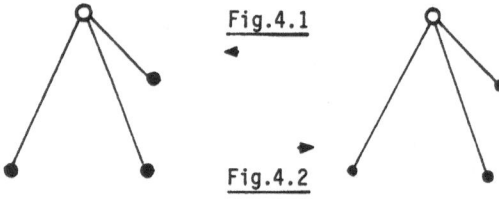

Fig.4.1. The structure of ammonia molecule

Fig.4.2. Inversion transition of the ammonia molecule

In the case of complicated macromolecular constructions, a situation may occur when a transition between two states differing from one another by their geometry and by their scheme of secondary bonds requires a strictly definite sequence of many small changes, every one of which can be described as a quantum jump. The realization of all of them, i.e., the transition between the initial and the final states, may take a long time. As a matter of fact, here lies the principal difference between the chemical behavior of low-molecular compounds and highly ordered macromolecular systems, i.e., constructions. Statistical physics of such kinetically nonequilibrium systems with rigid memory on different levels of organization has been developed since 1968 [4.11,12].

It is well known that protein reactions are, as a rule, accompanied by considerable conformational and configurational changes [4.2]. Conformation is determined by a scheme of secondary bonds. Configuration determines the spatial positions of the local equilibriums of atoms. Due to small perturbations, a configuration can be changed without any conformational transition taking place. Large enough configurational changes can lead to essential structural constraints and to a conformational transition.

In accordance with the above peculiarities of ordered protein structures, it is in the course of chemical reactions of proteins that we can expect the formation of specific long-living conformationally nonequilibrium states. A fast local chemical change (attachment of a ligand to the active center, ionization of acid or basic group, etc.) is accompanied by a fast vibrational relaxation ($\mathscr{T} \approx 10^{-12}$ s) of the active center and its immediate surroundings and takes place with the structure of the whole protein globule remaining practically unchanged. The subsequent changes are of relaxational nature and require a coordinated disruption and formation of a multitude of weak secondary bonds, i.e., the overcoming of a large entropy barrier. Conformational relaxation can last for microseconds, milliseconds, and even seconds. In this way, after a chemical transformation of the active center there appears a long-living specific out-of-equilibrium state: the active center is already changed, but the structure of the main volume of protein macromolecule remains the same although now it is out-of-equilibrium relative to the changed active center. A structural strain appears between the relaxed portion and the rest of the protein volume. This strain is slowly lifted in the course of protein conformational relaxation. In

74

the cases of sufficiently large proteins the resultant conformational transition can, probably, be completed only in the presence of this structural strain, only under the action of a *force*, which pushes the system towards a new conformational state, i.e., only after the appearance of a nonequilibrium state resulting from a fast local chemical transformation. In the absence of such a force the time required for the completion of conformational transition can be so great that this transition will be impossible to observe. This question will be considered further in Sect.4.3.

Two basic ideas of this monograph can be formulated as follows.

1) Conformational relaxation of proteins and their complexes after fast local chemical changes must be regarded as motion along a specific mechanical degree of freedom.

2) This conformational relaxation is essentially an elementary act of such processes as enzymatic catalysis and the energy coupling of intracellular chemical reactions.

The most convenient objects for the generation and study of conformationally out-of-equilibrium states are, probably, metal-containing proteins. Their active centers readily undergo fast chemical transformations. Their physical and chemical characteristics can be conveniently recorded by means of such methods as absorption spectrophotometry, circular dichroism, magnetic circular dichroism, Raman scattering, Mössbauer effect, EPR, etc.

There are two methods of generating and investigating metal-containing proteins in conformationally out-of-equilibrium states:

1) fast pulse technique at room temperatures (pulse photolysis, pulse radiolysis) which enables one to record the spectra and estimate the chemical reactivity of proteins at different times after fast (10^{-8} - 10^{-5} s) changes of their active centers and during subsequent conformational relaxation;

2) reduction or photolysis at low (4-77 K) temperatures, when the mobility of a macromolecule is essentially restricted. The latter method makes it possible to fix metal-containing proteins in kinetically stabilized conformationally out-of-equilibrium states and to study them with the help of the above-mentioned powerful methods.

The conformationally nonequilibrium states of metal-containing proteins were first investigated by means of the pulse technique at room temperatures. As early as 1959 GIBSON [4.13] studied carboxyhemoglobin using the pulse photolysis method, and concluded that carbon monoxide dissociation results in the appearance of nonequilibrium forms of reduced protein with enhanced reactivity. Later investigations showed that Gibson's results can be explained if we assume the photolytic formation of two equilibrium forms of so-called R and T hemoglobin conformers having a different number of CO groups attached to the protein. The use of nanosecond laser technique showed, however, that even in this case, short-living nonequilibrium forms of hemoglobin actually appear [4.14].

Significant progress in the study of dynamic and spectral properties of proteins in nonequilibrium states has been made with the help of pulse radiolysis technique. Metal ions in the active centers of metal-containing proteins are reduced by hydrated electrons formed during radiolysis within 10 μs. A fast reduction of proteins, whose equilibrium conformation and/or active center structure depend on the oxidation state of metal ions in their active centers, is, as a rule, followed by the appearance of the protein nonequilibrium forms with lifetimes from 10^{-5} to 1 s.

The experimentally observed spectral and dynamical characteristics of nonequilibrium states of metal-containing proteins will be described somewhat later. We shall first consider the data obtained with the help of low-temperature radiolysis of photolysis of metal-containing proteins in frozen solutions, suspensions of cells or subcellular particles, and in the whole tissues.

4.2.1 Nonequilibrium States of Metal-Containing Proteins Trapped at Low Temperatures in a Frozen Matrix

The method of low-temperature reduction was developed especially to obtain and trap the out-of-equilibrium forms of metal-containing proteins and to study their spectral and magnetic characteristics [4.15,16]. In this method the protein metal-containing prosthetic group is reduced in a frozen water solution at 77 K with thermalized electrons (e_t^-) obtained by radiolysis. The initial preparations were oxidized proteins whose active centers can exist both in oxidized and in reduced forms (hemoglobin, myoglobin, cytochromes, iron-sulfur proteins). The affective redox potential of e_t^- is so negative that they are able to reduce all the proteins studied. To make preparations glassified[1] at low temperatures, and also as protectors, small quantities of ethylene glycol, glycerine or sucrose can be added.

After the active center electron reduction a kinetically stabilized nonequilibrium state appears: the metal atom in the active center is reduced, but the stereochemistry of its immediate surroundings is changed to the extent permitted by the unchanged protein globule frozen into the matrix. The immediate surroundings of the active center undergo vibrational relaxation, but the spatial structure of the polypeptide chain remains the same as it was in the equilibrium oxidized protein. This new state of the active center must, however, correspond at equilibrium to the changed positions of the ligands and a changed conformation of the whole protein globule. The existence of a strain between the changed active center and the other parts of the macromolecule affects the spectral and magnetic characteristics of the active center. In this way the specific conformationally out-of-equilibrium states of metal-containing proteins appear: the metal ion is reduced, but the structure of the major part of protein globule corresponds to the oxidized state of this metal ion.

---

1 It means converted into the glassy, transparent solid state.

This method of generation, fixation and investigation of conformationally non-equilibrium states can be applied not only to the study of isolated proteins, but of subcellular organelles, cells and the whole tissues as well. Details of the experimental technique can be found in the articles cited. It is, however, necessary to describe here certain control experiments to make sure that the results obtained are trustworthy.

With an increase in temperature up to 140-200 K, relaxation of nonequilibrium proteins to their equilibrium states can be observed. The relaxation temperature depends both on the nature of protein and the matrix properties. The optical, magnetic, chemical (enzymatic included) properties of these equilibrium states obtained after relaxation are indistinguishable from those of equilibrium proteins reduced at room temperature with the help of conventional chemical methods. After the radiolysis of already reduced equilibrium proteins no changes in their optical, magnetic or chemical properties were recorded.

Nonequilibrium states of many metal-containing proteins, both isolated and located within intracellular membrane structures, have been studied by means of this method. Let us begin with proteins whose prostetic groups are iron-porphyrins, that is hemes. Spectral and magnetic properties of heme compounds are determined by both the oxidation and spin states of the iron atom. Four types of states are to be distinguished: Fe(II) high-spin (the number of unpaired electrons, $n = 4$), Fe(II) low-spin ($n = 0$), Fe(III) high-spin ($n = 5$), Fe(III) low-spin ($n = 1$). In the case of heme complexes the iron spin state is mainly determined by the nature of the fifth and sixth axial ligands. Four types of spectral and magnetic characteristics of the heme-containing proteins exist corresponding to the four types of electron states listed above. The splitting of iron d-orbital energy levels in ligand fields of different symmetry is schematically shown in Fig.4.3. As a rule, there exists a strict correlation between the position of iron atom relative to the porphyrin plane, its spin state and the nature of the axial ligands.

The presence of two strong (in terms of the ligand field theory) axial ligands leads to large splittings between the levels with different m_L values, and consequently, to a low-spin electron configuration. Strong ligands are imidazole, pyridine, CN^-, N_3^-, O_2, CO, methionine sulfur in cytochrome c, etc. The iron atom in these cases is either localized in the porphyrin plane (in ferro-derivatives) or slightly displaced relative to this plane (in ferri-derivatives). If the sixth axial ligand is weak (H_2O, F^-, etc.) or altogether absent, the iron atom has a high-spin electron configuration and is displaced from the porphyrin plane by 0.02-0.3 Å in ferri-derivatives, and by 0.5-0.7 Å in ferri-derivatives. Typical spectral characteristics of the above four types of iron-porphyrin complexes are shown in Fig.4.4.

Low-temperature reduction with thermalized electrons was applied to the generation and study of the nonequilibrium states of cytochrome c of animal and plant origin (horse heart, yeast, beef adrenal, chloroplasts of green plants, chromato-

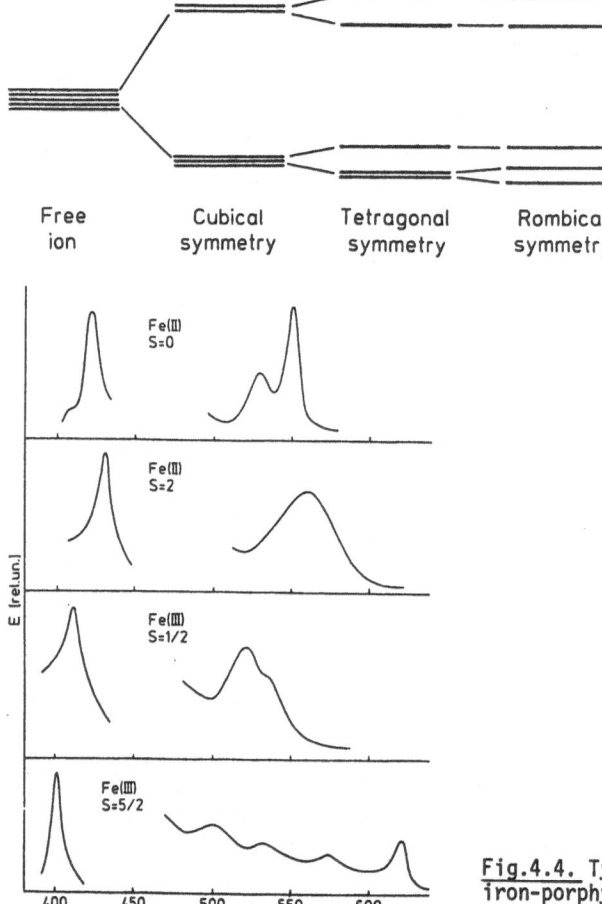

Fig.4.3. Splittings of the iron d-orbital energy levels in the ligand fields of different symmetry

| Free ion | Cubical symmetry | Tetragonal symmetry | Rombical symmetry |

Fe(II) S=0

Fe(II) S=2

Fe(III) S=1/2

Fe(III) S=5/2

E [rel.un.]

400   450   500   550   600
[nm]

Fig.4.4. Typical absorption spectra of iron-porphyrin complexes

phores of photosynthetic bacteria); cytochrome c complexes with low-molecular ligands (CN^-, F^-, N_3^-, imidazole); modified cytochrome c (carboxymethylated at methionine 80 and formylated at tryptophan 59). By means of absorption spectroscopy and MCD methods the nonequilibrium states of active centers differing by their absorption and MCD spectra from the corresponding equilibrium proteins were recorded in all cases. A detailed description of the experimental technique used can be found in the original papers [4.16-21]. Certain important results will be described here.

The differences between electron characteristics of active centers in the non-equilibrium forms of various cytochromes c and the corresponding equilibrium proteins are definitely trustworthy but rather small. These differences are usually confined to small shifts of absorption bonds, changes in their vibrational structure, in their relative intensities. Accordingly, these differences reflect the differences between the structures of active centers in the equilibrium reduced and equilibrium oxidized proteins. It is certainly very well known that in the case of cytochrome c, these differences are not so great [4.22,23]. It should be noted, at the same time, that the differences between spectral characteristics of equilibrium and nonequilibrium forms of cytochrome c depend essentially on the origin of the protein. For example, these differences are almost negligible for cytochrome c obtained from yeast, but are quite prononced for cytochrome c isolated from photosynthesizing plant cells. The differences are much more pronounced in the case of beef adrenal cytochrome c than in the case of horse heart cytochrome c. These examples could be continued.

According to the properties of nonequilibrium states fixed by low-temperature solutions, cytochrome c complexes with low-molecular ligands can be separated into two groups. The first group includes ferricytochrome c complexes which dissociate after reduction: the equilibrium ferro-cytochrome c is unable to form complexes with these ligands. These ligands are N_3^-, CN^-, imidazole, and in the case of carbocymethyl-ated cytochrome c, F^- ion. As a rule, nonequilibrium states of these complexes differ essentially in their spectral characteristics from the respective equilibrium forms. In this case the low-molecular ligand was found to remain in the coordination sphere of heme iron even after its reduction. Cytochrome c at moderate alkaline pH values ($9.0 \leq pH \leq 11.5$) can also be placed in this group. Under these conditions, axial ligands of the iron atom are his-18 and lys-79 in ferricytochrome c, but his-18 and met-80 in ferrocytochrome c. Therefore, reduction of ferricytochrome c in moderately alkaline frozen solutions with thermalized electrons leads to the formation of non-equilibrium states whose absorption and MCD spectra differ greatly from those of the equilibrium protein.

In cytochrome c complexes of the second group (they include cytochrome c itself at neutral pH values and imidazole complexes of carboxymethylated and formylated cytochrome c) the iron ligand sphere does not depend on the iron redox state, and the spectral characteristics of equilibrium and nonequilibrium states do not differ so greatly. Comparatively small differences of the active center electron character-istics in a conformationally nonequilibrium state have also been found for cytochrome b_5 reduced in the frozen water-glycerine matrix with thermalized electrons [4.24]. Cytochrome b_5 is a component of microsomal ETC.

Interesting results were obtained in the study of hemoglobin and myoglobin. A hemoglobin molecule has two "α" and two "β" subunits. The fifth, axial, ligand of an iron atom is the nitrogen atom in the imidazole group of the so-called proximal histidine residue. An important role in the formation of hemoglobin spatial structure is played by hydrogen bonds between the porphyrin propionic acid groups and amino acid residues of the polypeptide chain, as well as by a multitude of hydrophobic contacts between heme and different parts of protein globule. In the physiologically active ferro-forms, the sixth position in the iron coordination sphere is either vacant (deoxyhemoglobin) or occupied by O_2 (oxyhemoglobin). Besides oxygen, CO, NO and various alkylisocyanides (but not N_3^-, F^-, or imidazole) may be attached to the sixth position of Fe(II). In the ferrihemoglobin (also called "methemoglobin") the sixth, axial, ligand of Fe(III) is a water molecule (or OH^- at pH > 8.0). H_2O can be substituted by CN^-, N_3^-, imidazole.

Heme is situated in a hydrophobic cleft (the so-called heme pocket) between two helical regions of the polypeptide chain [4.25]. In the immediate neighborhood to the heme lies the so-called distal histidine, creating substantial steric hindrances for such ligands as CO and CN^- and forming hydrogen bonds with H_2O or N_3^- molecules in the ferrihemoglobin coordination sphere.

In equilibrium deoxyhemoglobin (ferrohemoglobin) the pentacoordinated Fe(II) is in a high-spin state (S = 2) and is displaced relative to the porphyrin plane in the direction of distal histidine by 0.60 Å in α-subunits and by 0.65 Å in β-subunits. The O_2,CO or NO binding is accompanied by heme iron transition into a low-spin state (S = 0) and by a sharp decrease (in the equilibrium form of the protein) of its distance from the porphyrin plane: for HbCO this distance is equal to 0.04 Å (in α-subunits) and to 0.22 Å (in β-subunits) [4.26]. At the same time the structures of active centers in the equilibrium *high-spin* ferrihemoglobin are similar to their structures in *low-spin* ferrohemoglobin: the iron atom is localized at 0.03 Å from the porphyrin plane in β-subunits and at ~0.2 Å—in α-subunits. A detailed theoretical discussion of these structural characteristics can be found in [4.27].

In the light of the aforesaid it is not surprising that after the reduction of ferrihemoglobin in a frozen matrix (pH 6.0-7.0) up to 80% of the protein exists in a nonequilibrium low-spin reduced form characterized by the "hemochromogenlike" absorption spectrum (Fig.4.5) and by the MCD spectrum typical for the low-spin heme compounds [4.15,28]. Similar effects are observed in the case of a high-spin ferrihemoglobin fluorine complex after low-temperature reduction [4.28]. After incubation at higher temperatures the electron characteristics of relaxed protein do not differ from those of equilibrium deoxyhemoglobin. The reduction of the azide, cyanide, imidazole, or OH^- (pH 8.5) ferrihemoglobin complexes leads to the formation, with a 100% yield, of nonequilibrium low-spin reduced forms of protein with absorption and MCD spectra of the hemochromogen type [4.28-31].

Analysis of these data shows that the formation of low-spin nonequilibrium forms of ferrohemoglobin and its complexes after low-temperature reduction of ferrihemoglobin and its complexes is caused by the preservation of axial ligands in the sixth coordination position of Fe(II). The role of such a ligand can be played by F^-, CN^-, N_3^-, imidazole, OH^-, and (in the case of ferrihemoglobin at pH 6.5-7.0) by distal histidin or an H_2O molecule which forms a hydrogen bond with distal histidine. Ligand dissociation is blocked by steric hindrances at the protein groups around the heme. The iron atom cannot move away from the porphyrin plane far enough to ensure the equilibrium high-spin state, and the electron configuration becomes a low-spin one. For ferrihemoglobin at pH 6.5-7.0 about 20% of protein reduced with thermalized electrons is in a high-spin state, the spectral and magnetic characteristics of which are quite similar to those of equilibrium ferrohemoglobin. Evidently, about 20% of equilibrium ferrihemoglobin in the initial solution has such structural organization of the polypeptide chain that there are no hindrances for ligand

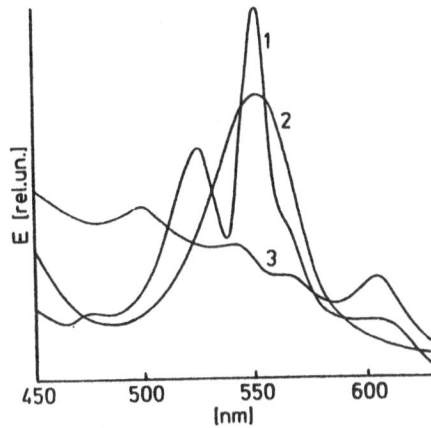

Fig.4.5. Absorption spectra of nonequilibrium ferrohemoglobin (1), equilibrium ferrohemoglobin (2), and equilibrium ferrihemoglobin (3) in frozen solutions at 77 K

expelling, and nothing prevents the formation of a high-spin state with spectral properties similar (but not necessarily identical) to those of equilibrium ferrohemoglobin. Almost the same results have been obtained with low-temperature reduction of ferrimyoglobin and its complexes [4.32].

In equilibrium horseradish peroxidase, Fe(II) exists in a pentacoordinated high-spin state. Its reduction with thermalized electrons in a frozen matrix results in the formation of two nonequilibrium conformers: low-spin and high-spin ones [4.33,34]. As distinct from hemoglobin, the absorption and MCD spectra of peroxidase nonequilibrium high-spin forms differ essentially from those of equilibrium protein. In the initial ferriperoxidase solution there seem to exist two conformers differing structurally from one another, as well as from the equilibrium ferroperoxidase. In a conformer responsible for the formation of the high-spin nonequilibrium state, Fe(III) ion in the active center either already exists in a pentacoordinated state or the expelling of the sixth ligand is not hindered sterically. In the nonequilibrium F^-, N_3^- and CN^- complexes of ferroperoxidase, formed by the reduction of the corresponding ferriprotein complexes, heme is in a low-spin state. These complexes give absorption and MCD spectra of the hemochromogen type.

As already mentioned, the conformationally nonequilibrium states of hemoglobin, myoglobin, peroxidase and their complexes relax with temperature increase to the respective equilibrium states even before the matrix is defrosted. It was possible to establish in certain cases the sequence of the structural changes taking place. It has been found, e.g., that before the axial ligand is expelled from the iron atom coordination sphere its orientation changes, and the iron shifts from the porphyrine plane. Relaxation of the protein nonequilibrium forms can proceed via several intermediate states. Ligand dissociation, for example, does not always lead to the formation of the final equilibrium state. Conformationally out-of-equilibrium states were also recorded after low-temperature reduction of microsomal cytochrome P450 [4.35, 36].

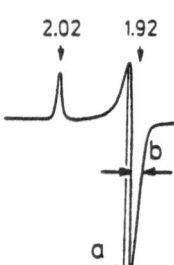

Fig.4.6. Binuclear Fe-S-cluster and EPR signal of adreno-doxin in a frozen solution at 77 K

Nonequilibrium states arising after the active center reduction with thermalized electrons have been found both in the experiments with purified cytochrome-c-oxidase preparations, and in similar experiments with whole mitochondria or submitochondrial particles [4.37]. In a conformationally nonequilibrium state the absorption bands of cytochrome-c-oxidase are shifted by 3-4 nm towards the red. These spectral changes are mainly caused by the cytochrome a forming a nonequilibrium state: the spectral changes of cytochrome a_3 are much less pronounced.

The appearance of structurally nonequilibrium states after low-temperature reduction is not the prerogative of hemoproteins. Substantial effects have been observed in the case of iron sulfur proteins with Fe-S-clusters as active centers (Chap.2). These proteins were studied in isolated form as well as a part of submitochondrial particles, mitochondria, chloroplasts or the whole tissues, mainly with the help of the EPR method. Let us consider as an example the origin of the EPR spectrum for a binuclear Fe-S-cluster (Fig.4.6). In the oxidized form of the cluster both iron ions are in high-spin states ($S_1 = S_2 = 5/2$). Due to antiferromagnetic coupling the system's ground-state summary spin in zero ($S = 0$), the protein is diamagnetic and does not give an EPR signal. After one-electron reduction ($S_1 = 5/2$, $S_2 = 2$) in the ground state, antiferromagnetic coupling leads to $S = 1/2$ and to an EPR signal (Fig.4.6). The first excited state with $S = 3/2$ lies low enough ($\Delta E_{1/2,3/2} \approx 10^2$ cm^{-1}), and spin-lattice relaxation is realized via this excited state according to Orbach's mechanism [4.38]. The exchange integral determines $\Delta E_{1/2,3/2}$ and is also extremely sensitive to the changes of cluster structure (the exchange integral strongly depends on the iron-iron distance). The most sensitive structural text for Fe-S-proteins is, therefore, the spin-lattice relaxation time (T_1). Changes in T_1 reveal themselves in the width and shape of the EPR signal components and in their microwave saturation properties.

An Fe-S-protein, adrenodoxin, has been studied in [4.39]. The equilibrium reduced protein gives an EPR signal (Fig.4.6), and the $g_\perp$ component has the following parameters: a = 20 Gauss, b = 16 Gauss. In a conformationally nonequilibrium state obtained by the low-temperature reduction of oxidized adrenodoxin with thermalized electrons in a frozen matrix, the EPR signal is markedly broadened (a = 32 Gauss, b = 31 Gauss). It means that T_1 of the nonequilibrium center is decreased in comparison with T_1 value for equilibrium protein: the EPR signal components can be treated as homogeneously broadened lines (theoretical principles of EPR spectroscopy can be found, e.g., in [4.40]).

In the realm of Orbach's mechanism the decreased T_1 value corresponds to the decreased energy gap between the ground and the first excited spin states, i.e., to the decreased exchange integral absolute value, and, consequently, to the increased iron-iron distance in an Fe-S-cluster. It was found that iron reduction leads in the end to the cluster expansion [4.41]. This means that in the course of protein reduction the iron-iron distance at first increases, passes through a maximum (probably after vibrational relaxation) and then decreases again during conformational relaxation to the equilibrium value of reduced protein. In a frozen matrix this last stage cannot take place. The nonequilibrium form of adrenodoxin in a frozen water-organic matrix relaxes to its equilibrium form at temperatures between 160-210 K (Table 4.1).

Table 4.1. Changes in the values of parameters for the $g_\perp$ component of FPR signal of adrenodoxin reduced with thermalized electrons in a frozen solution depending on 15' incubation temperature. Measurements at 77 K (Fig.4.6)

T_{inc} [K]	a [Gauss]	b [Gauss]
77	32	31
143	32	31
157	32	28
178	28	23
195	25	21
210	20	17
290	20	16

The lines in EPR spectra of mitochondria, chloroplasts and tissues generally overlap to a considerable extent, and it is difficult to measure the width and shape parameters of an individual line. In our laboratory the method of "I(T) curves" has been proposed, with the help of which structural differences between Fe-S centers in the equilibrium and the nonequilibrium states can be detected.

Figure 4.7 shows an equilibrium reduced pea ferredoxin EPR spectrum recorded at 20 K. The differential intensity (I) of the signal can be measured using either the maximum at $g_z \approx 2.05$ or the minimum at $g_x \approx 1.89$. Figure 4.8 shows I(T) curves of the equilibrium (reduced with hydrosulfite at room temperature) and the nonequilibrium (reduced with thermalized electrons at 77 K) ferredoxin [4.42,43]. A characteristic feature of the conformationally out-of-equilibrium states of all the Fe-S-proteins studied is the shift of I(T) curves towards low temperature (as compared with the corresponding equilibrium protein). At a fixed microwave power this shift indicates an enhanced spin-lattice interaction in a nonequilibrium state. As already mentioned, enhanced spin-lattice interaction (i.e., decreased T_1) implies, in this case, an in-

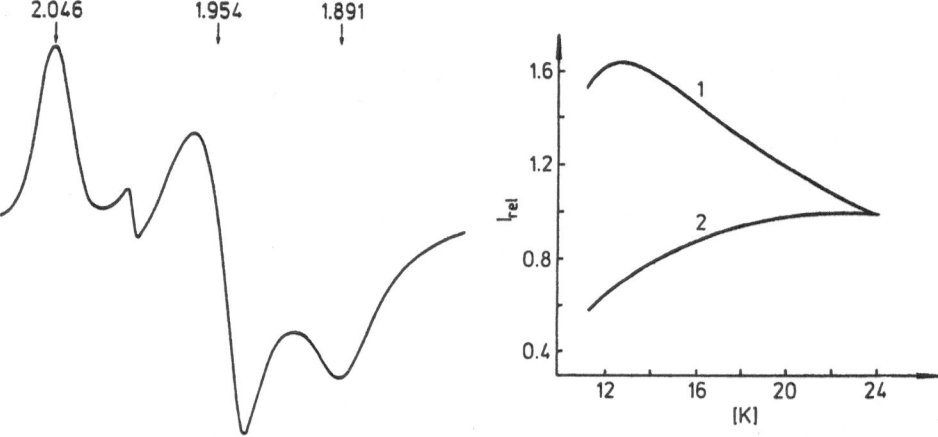

Fig.4.7. EPR spectrum of the equilibrium reduced pea ferrodoxin at 20 K. The principal g-tensor values are indicated. The signal at g ≈ 2.00 is due to a free radical admixture

Fig.4.8. I(T) curves (g_x component) of the conformationally nonequilibrium (1) and equilibrium (2) pea ferredoxin

creased iron-iron distance. The analysis of I(T) curves permits in certain cases the estimation of the exchange integral values in equilibrium and nonequilibrium states [4.44,45].

As an example of the recording of Fe-S-centers in conformationally nonequilibrium states within membrane structures let us consider the results obtained in the study of the N-2 center in mitochondrial ETC. Figure 4.9 shows the EPR spectrum of Fe-S centers in mitochondria at 13 K. The $g_{\parallel}$ and $g_{\perp}$ components of N-2 EPR signal are indicated. In Fig.4.10 I(T) curves for $g_{\perp}$ component in equilibrium and nonequilibrium states are shown.

Experiments were carried out as follows. Mitochondria uncoupled by aging[2] were subjected to low-temperature oxidation at ~10°C. Under these conditions the access of substrates into mitochondria is hindered, but diffusion of molecular oxygen is, nevertheless, quite intensive. All the ETC carriers, therefore, become oxidized. After that the mitochondrial preparation was subjected to the action of thermalized electrons at 77 K, and all carriers became reduced. The I(T) curve for the $g_{\perp}$ component of the N-2 center obtained with this preparation is shown in Fig.4.10 (curve 1). Curve 2 was recorded for the same preparation after its relaxation at 210 K. The same curve can be obtained if we reduce ETC carriers with the substrate excess before freezing.

---

2 N-2 center participates directly in the coupling between respiration and phosphoryl- ation in mitochondrial ETC, and its state depends essentially on the mitochondrial phosphorylating activity prior to freezing. Therefore, I shall describe here the results obtained with uncoupled mitochondria which fully retain their respiration activity but are unable to carry out oxidative phosphorylation.

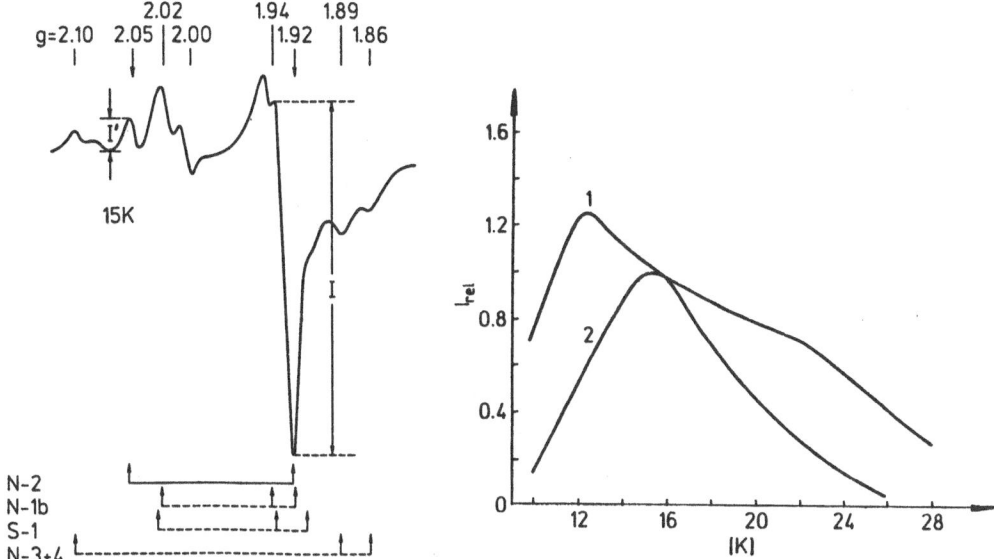

Fig.4.9. The EPR spectrum of beef heart mitochondria at 13 K. I and I' indicate intensities of the $q_\perp$ and $g_\parallel$ components of the EPR signal of N-2 center, respectively

Fig.4.10. I(T) curves ($g_\perp$ component) for N-2 center in beef heart mitochondria in the nonequilibrium (1) and equilibrium (2) states

The results described in this section show that the fixation of a protein con-formationally nonequilibrium state in a rigid matrix makes it possible to separate the local electron reduction of the active center from subsequent protein conforma-tional relaxation. Structural and electron properties of a metal-containing active center in a nonequilibrium state can differ substantially from those for the equi-librium protein.

4.2.2 Nonequilibrium States of Metal-Containing Proteins Recorded at Room Temperatures, and Kinetics of Their Relaxation

The conformationally nonequilibrium states of metal-containing proteins at room temperatures can be formed with the help of the following methods

1) Reduction of the metal ion in the active center by microsecond pulses of fast electrons or by aromatic radical anions with high negative values of redox potentials (radical anions are obtained by photoreduction of corresponding compounds using light flashes of short duration).

2) Flash photolysis of heme-containing protein complexes with CO or CN^-.

3) Jump-like changes in pH or ionic strength values by the stop flow method.

The spectral characteristics and relaxation kinetics of protein nonequilibrium states at room temperatures were most comprehensively studied with proteins whose active centers had been reduced by hydrated electrons. The first results were pub-lished in the early seventies [4.46-53]. Systematic investigations of metal-containing proteins in conformationally nonequilibrium states at room temperatures have recently been carried out in our laboratory [4.54-58]. In all the proteins studied, electron reduction of the active center leads to the appearance of nonequilibrium states, whose spectral characteristics at the initial steps of relaxation seemed to be very similar to the characteristics of the corresponding proteins after the reduction of their active centers in a frozen matrix at low temperatures. (It was possible to record spectra during relaxation beginning about 10 μs after the transformation of the active center.) The physical and chemical properties of these proteins after their relaxation are indistinguishable from those of the corresponding proteins reduced under equilibrium conditions. The treatment of previously reduced proteins with hydrated electrons does not change their characteristics.

Relaxation always takes place as a multistage process the rate constants of individual stages (λ_i) generally differing to a very considerable degree. Some results are presented in Table 4.2.

Let us consider the data concerning individual proteins. Ferricytochrome c reduction in water solutions at pH 6.0-7.5 (phosphate buffer) by microsecond electron pulses leads to the formation of ferrocytochrome c in a nonequilibrium state with changed spectral characteristics. (The rate constant of cytochrome c reduction by solvated electrons at room temperature is 5×10^{10} mole^{-1}s^{-1}). Kinetics of the transition to the final equilibrium state can be described by the sum of three exponents with λ_i from 10^2 to 10^4 s^{-1}. At pH > 9.0 or in the presence of tertiary butyl alcohol (2.5 M), an additional slow stage appears with $\lambda_4 \approx$ 2-7 s^{-1}. Under these conditions the sixth axial ligand in ferrocytochrome c is methionine 80, and in ferricytochrome c, not methionine 80 but most probably lysine 79 [4.59]. This additional slow stage can, therefore, be interpreted as a conformational transition in the course of which methionine 80 returns to the Fe(II) coordination sphere. A similar phenomenon is observed after the reduction of ferricytochrome c complexes with CN$^-$ or N$_3^-$ ions. Removal of cyanide ion from the Fe(II) coordination sphere proceeds extremely slowly.

Long-living conformationally nonequilibrium states appear not only after the reduction of the cytochrome c active center, but after its oxidation as well. For instance, a sufficiently fast oxidation of equilibrium ferrocytochrome c in solution at pH 10.5 by ferricyanide leads under these conditions to the formation of nonequilibrium ferricytochrome c with absorption band at 695 nm. This band indicates that methionine 80 is in the Fe(II) coordination sphere. Relaxation to equilibrium state accompanied by the substitution of lysine 79 for methionine 80 and the disappearance of the 695 nm band has the rate constant of ~0.8 s^{-1}.

After ferrihemoglobin reduction in water solution at neutral pH values nonequilibrium states appear during the relaxation of which four stages have been recorded. About 40% of hemes are in a nonequilibrium low-spin 20 μs after the reduction, and 60% in a nonequilibrium high-spin state (Fig.4.11). The low-spin state passes into the nonequilibrium high-spin state, and the latter relaxes rather slowly via several

Table 4.2. Relaxation rate constants (λ_i) for some reduced nonequilibrium iron-containing proteins at 20°C

No.	Protein	$\lambda_i \ [s^{-1}]$
1.	Cytochrome c(pH 7.5)	$\lambda_1 = 2 \times 10^4$ $\lambda_2 = 2 \times 10^3$ $\lambda_3 = 80$
2.	Cytochrome c(pH 10.6)	$\lambda_1 = 2 \times 10^4$ $\lambda_2 = 2 \times 10^3$ $\lambda_3 = 3$
3.	Cytochrome c(pH 7.2 2.5 M tret. Butyl alcohol)	$\lambda_1 = 8 \times 10^3$ $\lambda_2 = 2 \times 10^3$ $\lambda_3 = 5$
4.	Carboxymethylated cytochrome c(pH 7.3)	$\lambda_1 = 8 \times 10^4$ $\lambda_2 = 2 \times 10^3$ $\lambda_3 = 15$
5.	Formylated Cytochrome c(pH 7.3)	$\lambda_1 = 3 \times 10^4$ $\lambda_2 = 6 \times 10^3$ $\lambda_3 = 5$
6.	Cytochrome c · N_3^-	$\lambda_1 = 2 \times 10^4$ $\lambda_2 = 1 \times 10^3$
7.	Cytochrome c · CN^-(pH 7.3)	$\lambda_1 = 2 \times 10^4$ $\lambda_2 = 1 \times 10^{-2}$
8.	Hemoglobin (pH 9.2)	$\lambda_1 = 9 \times 10^3$ $\lambda_2 = 2 \times 10^3$ $\lambda_3 = 3 \times 10^2$
9.	Myoglobin (pH 7.4)	$\lambda_1 = 1.5 \times 10^4$ $\lambda_2 = 3 \times 10^3$ $\lambda_3 = 6 \times 10^2$ $\lambda_4 = 20$
10.	Hemoglobin · N_3^-(pH 7.4)	$\lambda_1 = 6 \times 10^4$ $\lambda_2 = 6 \times 10^3$
11.	Hemoglobin · CN^-(pH 7.4)	$\lambda_1 = 1 \times 10^4$ $\lambda_2 = 4 \times 10^3$ $\lambda_3 = 5 \times 10^{-2}$
12.	Myoglobin · N_3^- (pH 7.4)	$\lambda_1 = 5 \times 10^4$ $\lambda_2 = 2 \times 10^3$
13.	Myoglobin · CN^-(pH 6.1)	$\lambda_1 = 8 \times 10^4$ $\lambda_2 = 8 \times 10^3$ $\lambda_3 = 5 \times 10^{-2}$
14.	Ferredoxin (pH 7.2)	$\lambda_1 = 7 \times 10^3$ $\lambda_2 = 6$

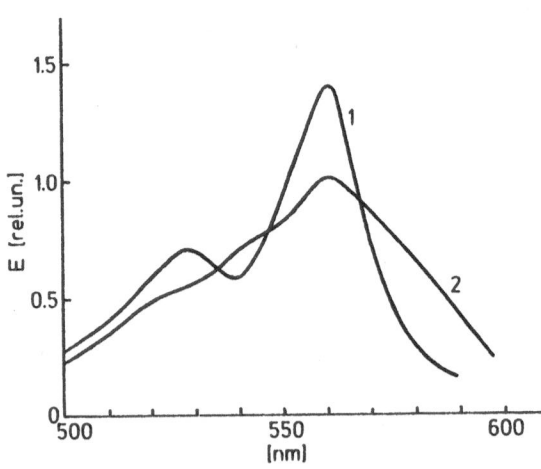

Fig.4.11. Absorption spectra of hemoglobin in solution after pulse reduction: after 20 μs (1) and after 800 μs (2)

stages to equilibrium. Similar effects have been also observed for myoglobin. Prolonged conformational relaxation is not a specific property of hemoproteins. Complex relaxation kinetics have been recorded after the pulse reduction of a typical Fe-S-protein-ferredoxin.

Hydrated electrons (e_h^-) are the most convenient reducing agents to use in studying the protein conformationally nonequilibrium states formed after reduction of protein active centers. This is due to the low value of the e_h^- redox potential $E^0 = -2.6$ V), which makes it possible to reduce the oxidized active centers of any metal-containing protein in any conformational state with diffusion rate constants exceeding 10^{10} mol^{-1}s^{-1}. The reduction rates of any metal-containing proteins with e_h^- are, as a rule, much higher than the rates of conformational relaxation. This facilitates their study.

When other less active reducing agents are used the recording of the intermediate nonequilibrium states depends on the experimental conditions and the electron donor properties. Data are, however, available showing that similar conformationally nonequilibrium states appear in the course of the usual chemical reduction of metal-containing proteins [4.60-64]. Naturally, because of a comparatively long duration of the reduction process, it is possible, in this case, to record only the most slowly relaxing nonequilibrium states at the final stages of conformational relaxation.

Conformationally out-of-equilibrium states have been observed not only in the course of redox transformations but also after photodissociation of hemoprotein complexes with CO and CN$^-$. It was shown with the help of the pico- and nanosecond laser techniques that hemoglobin and myoglobin formed after photodissociation of their carboxycomplexes differ slightly (by their absorption and Raman spectra) from the corresponding equilibrium proteins [4.14,65-67]. The lifetime of these states in solution does not exceed 10^{-7} s.

Much more long-living nonequilibrium states have been observed after the photodissociation of the complexes of cytochrome c and its derivatives, cytochrome P-450 and peroxidase with CO and CN⁻ [4.68-72]. Nonequilibrium states do not always become apparent through their spectral properties. The absorption spectra of horse radish peroxidase in equilibrium state and immediately after the pulse photodissociation of its cyanide complex practically coincide, although the rate constant of peroxidase and CN⁻ recombination during the first 100 ms after dissociation is considerably decreased as compared with the corresponding rate constant of CN⁻ binding to the equilibrium protein. In Sect.4.2.3 we shall discuss in greater detail the chemical properties of proteins in conformationally nonequilibrium states.

Rather slowly relaxing nonequilibrium states of cytochrome c have been recorded after fast changes of ionic strength [4.71] and pH [4.72]. In solutions with a low ionic strength the ferricytochrome c active center has a high-spin ground state, and in solutions with a high ionic strength the ground state turns out to be low-spin. Conformational relaxation is realized via several stages with characteristic time $T_{1/2} \approx 10^{-1}$ s. After a sufficiently fast pH decrease the transition from a low-spin (nonequilibrium under these conditions) to a high-spin state can be observed. Conformational relaxation consists of at least four stages with characteristic times from 10^{-3} to 10 s. In the course of relaxation the protein passes through nonequilibrium states which cannot be realized under equilibrium conditions at any intermediate pH values.

4.2.3 Chemical Properties of Metalloproteins in Conformationally Nonequilibrium States

Sufficiently long lifetimes of metal-containing proteins in conformationally nonequilibrium states make it possible to estimate their chemical reactivitiy. This has been done by measuring the rate constants of certain specific reactions of these proteins immediately after the generation of their nonequilibrium states and in the course of their conformational relaxation.

The following reactions were studied in detail: oxidation of reduced cytochrome c, hemoglobin, myoglobin, ferredoxin and their derivatives with potassium ferricyanide and plastocyanine; oxidation of myoglobin, hemoglobin and ferredoxin with ferricytochrome c; oxidation of cytochrome c with cytochrome-c-oxidase; interaction of hemoglobin, myoglobin and their complexes with O_2; attachment of CO (or recombination with CO after pulse photodissociation) to hemoglobin, myoglobin, cytochrome P-450, peroxidase, cytochrome c and its carboxymethylated derivative.

The main result of these kinetic studies is that the rate constants of the reactions of proteins in out-of-equilibrium states may differ essentially (up to 10^3 times) from those of the same proteins in equilibrium states. The reactivity of nonequilibrium proteins is usually enhanced, but the opposite effects can also be observed. Rate constants change in the course of relaxation to the values characteristic for equilibrium proteins. During conformational relaxation and in the presence of an excess concentration of one of the reagents, the reaction kinetics can be formally represented by a superposition of several reactions of pseudo-first order. It often happens that the duration of the first fast reaction stages of nonequilibrium heme-containing proteins (i.e., the duration of the stages with given intermediate rate constant values) is close to that of the initial stages of structural relaxation initiated by the active center reduction. In some cases, especially when both reagents are proteins, reactivity continues to change when the spectroscopically recorded structural changes no longer take place. If the reaction duration is appreciably longer than the duration of conformational relaxation, the rate constant value of the slowest stage coincides with that for the equilibrium protein. The data on the reactivity of nonequilibrium proteins obtained in our laboratory can be found in [4.54-56,61,69,70,73-75].

In all the protein reactions studied two processes are simultaneously taking place: the investigated chemical reaction (e.g., metal ion oxidation or CO binding) and protein conformational relaxation. Protein molecules at different relaxation stages are in fact different molecules with different electron structures of their active centers and different chemical properties. This produces unexpected effects generally unobserved in the reactions of low-molecular compounds. The measured rate constants of many chemical reactions prove to be dependent on the absolute velocity of the process determined by the concentrations of reagents. This can be observed, e.g., in the case of reactions of nonequilibrium myoglobin or hemoglobin with ferricyanide or oxygen. An increase in the low-molecular reagent concentration leads to an increase in the measured rate constant. Evidently, with the increase of the absolute rate of reaction a greater part of it can take place at the earlier stages of conformational relaxation when the protein is more active. If a reaction is sufficiently fast the protein will react practically all the time in a conformationally out-of-equilibrium state. These effects seem to explain the experiments showing that the calculated rate constants of electron transfer between the components of mitochondrial ETC increase with increasing respiration velocity [4.76].

Reaction rates of nonequilibrium proteins often have unusual temperature dependences. The initial configuration and the rate of conformational relaxation of proteins are, as a rule, much more sensitive to temperature changes than the rate of the studied reaction. Therefore, e.g., with decreasing temperature conformational relaxation slows down and a greater part of the reaction has enough time to be completed within the first stages of relaxation when the protein reactivity is maximal. The measured reaction rate will in this case increase in apparent contradiction to the Arrhenius law.

The reactivity of proteins in nonequilibrium states depends essentially on the reaction type. Let us take, e.g., myoglobin and hemoglobin in nonequilibrium states generated by pulse reduction of the corresponding ferriforms. Their reactivities in oxidation reactions are radically enhanced as compared with those of equilibrium reduced proteins. At the same time, in the reaction of CO or O_2 binding, the equili-

brium and the nonequilibrium forms of these proteins have practically the same reactivities. The rate of CO and O_2 binding to the majority of hemoproteins seems to
be determined not by the Fe(II) spin state, but by the Fe(II) accessibility, i.e.,
by steric hindrances along the path of CO or O_2 molecules to iron in the active
center. The equilibrium configurations of ferrihemoglobin and ferrimyoglobin globules (which do not exhibit any substantial changes during the first stages of conformational relaxation after pulse reduction) probably do not differ in this respect
from the equilibrium configurations of ferrohemoglobin and ferromyoglobin.

On the other hand, the rate constant of CO binding to the nonequilibrium ferrocytochrome c (obtained by the method of pulse reduction) is by about three orders
of magnitude higher than that for the equilibrium protein [4.61]. This can be explained by the fact that the active center of cytochrome c is more "open" in the
equilibrium oxidized state of this protein than in its equilibrium reduced state.

The formation of nonequilibrium states with decreased reactivities was recorded
in the studies of CO recombination with peroxidase [4.55] and cytochrome P-450 [4.69]
after flash photolysis of their carboxyderivatives. In a study of CO binding to
hemoglobin and of carboxyhemoglobin dissociation at low CO concentrations[3] under
conditions of continuous photolytic illumination [4.77-81], the authors were compelled to postulate the formation of several intermediate nonequilibrium hemoglobin
states. The rather peculiar kinetic effects observed during the transitions of the
system from the equilibrium dark state to the steady light state and backwards could
not be explained within the realm of classical concepts. Their interpretation was,
however, quite natural if one assumes that during the above-mentioned transitions
there appear long-living forms of carboxycomplexes with increased rate constants of
CO dissociation.

Especially interesting are the data concerning changes in the chemical properties
of metal-containing proteins in mitochondrial ETC. After fast pulse reoxidation of
reduced cytochrome-c-oxidase with molecular oxygen, the so-called pulse cytochrome-
c-oxidase appears that oxidizes ferrocytochrome c several times faster than an equilibrium "rest" protein [4.82]. This nonequilibrium cytochrome-c-oxidase differs from
the equilibrium protein not only in its reactivity but by its optical and EPR spectra
as well [4.83].

It was already noted that iron-sulfur center N-2 in the NADH dehydrogenase complex
of mitochondrial ETC can exist in an equilibrium and a conformationally nonequilibrium form. The reactivity of N-2 center in the nonequilibrium state is higher than
in the equilibrium state: its low-temperature oxidation with molecular oxygen proceeds much faster [4.84].

---

3 When CO concentration is lower than that of hemoglobin, practically no CO molecules exist in the conditions of dark equilibrium. This makes the kinetic analysis
  of the processes in question much easier.

The recording of proteins in conformationally out-of-equilibrium states described in this chapter requires sophisticated experimental technique: low temperatures to slow down the relaxation, pulse methods to realize simultaneously the local chemical changes in a considerable fraction of protein molecules within the sample. It is quite clear, however, that similar events must take place with individual protein molecules in the course of their functioning at room temperatures without forced synchronization. Thus, the data presented in this chapter prove that many protein reactions do actually proceed in two principal steps: a fast local change and a subsequent slow relaxation in the course of which protein molecules pass through a series of essentially nonequilibrium states.

4.3 Nonequilibrium Mixture of Molecules or Nonequilibrium Molecules?

When discussing the problems of nonequilibrium states and their relaxation dealt with in this chapter, one is usually faced with two extreme contradicting points of view. The first one can be called the orthodox statistical approach, according to which in this case we are dealing not with the appearance of principally new states of molecules but with a redistribution of sample molecules among the allowed conformer types, with the creation of a nonequilibrium mixture of these same conformers, which are *in principle* present even at equilibrium but cannot be observed because of the smallness of their concentrations. According to this point of view, conformational relaxation is a process of gradual changes in the composition of conformer mixture due to "momentary" transitions of individual conformers between discrete conformational states up to the point when equilibrium distribution is reached. This point of view implies, e.g., that low-spin state recorded after iron reduction in ferrihemoglobin, where the active center is reduced but most of the protein preserves the characteristic structure of an oxidized form, reflects only the enrichment of the mixture with a conformer that is also present (although at a low concentration) in the equilibrium ferrohemoglobin solution.

According to the second extreme point of view that can be called the orthodox mechanical approach, we observe in this case the appearance of new states that do not exist in dynamical equilibrium with the conformers present under equilibrium conditions. In the course of conformational relaxation every molecule passes through states that *in principle* are not realized in the conditions of thermodynamic equilibrium.

I think that this discussion is purely semantic, and the contraposition of the two extreme points of view is meaningless. As a matter of fact the true subject of this discussion is the validity of the ergodic hypothesis for complex molecules, when an individual molecule can be regarded as a statistical system with a large number of almost isoenergetic states and a "memory", i.e., large enough kinetic (enthalpy as well as entropy) barriers separating different regions of the phase space [4.2,11]. To be sure, every ferrohemoglobin molecule in an equilibrium system can some time assume a state in which most of the protein globule has the structure of equilibrium ferrihemoglobin although the iron ion in the active center remains reduced and its immediate surroundings are at local equilibrium. This means that in a thermodynamically equilibrium mixture composed of *a large enough number* of ferrohemoglobin molecules we can find a certain number of molecules in the above-mentioned state. The first point of view is thus *in principle* true.

On the other hand, however, in the case of a sufficiently complex system this "some time" may correspond to such a long time interval (even much longer than the lifetime of the universe) that at any reasonable volume of our thermodynamically equilibrium mixture of conformers and during any reasonable time interval such a transition would not take place even once in a protein molecule. The second point of view is thus *in principle* also correct.

The relaxation of a nonequilibrium state *specially prepared* by means of directed local action can *in principle* be described with the same justification either as a sequence of elementary acts every one of which proceeds reversibly and transfers a molecule into a new conformational state (this new "conformer" may differ from the old one, e.g., only in the turn of one link around a single bond), or as continuous motion along a specific mechanical degree of freedom. In the case of simple enough molecules the first approach is more preferable. During the relaxation of complex protein molecules the elementary acts are: turns around single covalent bonds, changes of individual valence angles, ruptures and formations of hydrogen bonds, etc., which are to be performed in a strictly definite sequence relaxed by means of a multitude of try-and-error acts. When the required number of these elementary acts is large enough the description of a relaxation process in the realm of the first approach is inconvenient, practically impossible and, therefore, meaningless, as, e.g., the microscopic description of any macroscopic mechanical motion.

5. Conformational Relaxation as the Elementary Act of Bioenergetic Processes

In the preceding chapters the main distinctive features of bioenergetic processes and the most important characteristics of the relevant macromolecular structures taking part in these processes have been considered. Based on this consideration the following conclusions can be inferred.

1) Energy transduction in living systems is realized according to the principle of energy coupling between the energy-donating and the energy-accepting chemical reactions.

2) This coupling is realized by means of special macromolecular devices, machines which perform energy transduction practically independently of each other.

3) An elementary act of any bioenergetic process proceeds according to the following scheme: fast local perturbation and subsequent slow conformational relaxation in the course of which coupling between energy-donating and energy-accepting reactions is actualized.

4) The conformational relaxation of a molecular machine is to be regarded as mechanical motion. The functioning of machines of molecular dimensions always includes a stage of thermodynamically irreversible conformational relaxation. This stage represents the working stroke of a molecular machine.

Two last conclusions are, as a matter of fact, the formulation of the relaxation concept in bioenergetics [5.1]. In this chapter the main bioenergetic processes will be discussed on the basis of this concept. As in Chap.2 we begin with muscle contraction.

5.1 Muscle Contraction

In recent years papers have begun to appear in the literature, the authors of which use mechanical concepts when discussing the elementary steps of muscle contraction (or of biological motility in general). Various concrete devices and various chemical-mechanical schemes are postulated. For instance, in [5.2] it is assumed that gradual ATP changes during hydrolysis at the substrate binding site in the myosin S-1 fragment lead allosterically to a change in the affinity between myosin and actin in another part of S-1 fragment, as well as to a change in the dip angle of actin monomer relative to the anisotropic S-1 frame. The sequence of these gradual changes leads to a relative displacement of myosin and actin filaments, i.e., to mechanical motion.

An even more minute description of the chemimechanical energy transduction during muscle contraction is given in a model proposed by LEVY et al. [5.3]. The postulated chemical scheme of the process can be written as follows

$$\underbrace{AM}_{a} + MgATP \longrightarrow \underbrace{AM^* \cdot MgATP}_{b} \longrightarrow A + \underbrace{M^* \cdot MgATP}_{c} \longrightarrow$$

$$\longrightarrow A + \underbrace{M^* \cdot ADP \cdot Mg \cdot P_i}_{d,e} \longrightarrow \underbrace{AM^{**} \cdot ADP \cdot Mg \cdot P_i}_{f} \longrightarrow \underbrace{AM^* \cdot ADP \cdot Mg \cdot P_i}_{g} \longrightarrow$$

$$\longrightarrow \underbrace{AM^*}_{h} + P_i + MgADP \longrightarrow \underbrace{AM}_{a} + P_i + MgADP \qquad . \quad (5.1)$$

Here A is actin, M is myosin, AM is actomysin, and the asterisks designate different conformationally changed states of the myosin molecule.

The mechanical scheme of the energy transduction process proposed in [5.3] is shown in Fig.5.1. In state (a) the distal end of S-1 is attached to one of the actin monomers. The "hinge" is completely relaxed, there is no conformational strain. The binding of MgATP molecule to the S-1 active center leads, according to KOSHLAND's induced fit mechanism [5.4], to conformational changes ensuring a gradual strengthening of the MgATP — (S-1) bond and a gradual increase of constrains in the region of A-(S-1) bond, state (b). This leads ultimately to the dissociation of the S-1 fragment from actin, state (c). ATP hydrolysis, as in the original Limn-Taylor and Bukatina-Deshcherevsky schemes (Chap.2), takes place after the rupture of the actin-myosin bond. The primary and, according to [5.3], reversible energy transduction process takes place during the (c) → (d) → (e) transition. The energy liberated during hydrolysis of ATP high-energy bond is trapped in the hinge region, the angle between S-1 and S-2 changes markedly, and the S-1 active center near the hinge approaches the actin filament, state (e). The liberated energy is now concentrated at the hinge. Elastic deformation has occurred, and the hinge "spring" is ready to perform work as soon as the actin binding initiates the next stage of the enzymatic process, state (f). Interaction between S-1 and actin is now localized near the S-1 catalytic center and the·hinge. This interaction now releases the hinge "latch", and, consequently, a conformational relaxation can take place, states (g) and (h). This relaxation consists in a change of the angle between S-1 and S-2, in the course of which the products of hydrolysis are liberated, the bond between S-1 and the other actin monomer is formed, and the actin filament is pushed relative to that of myosin. During this process the mechanical effort arises only at the hinge region

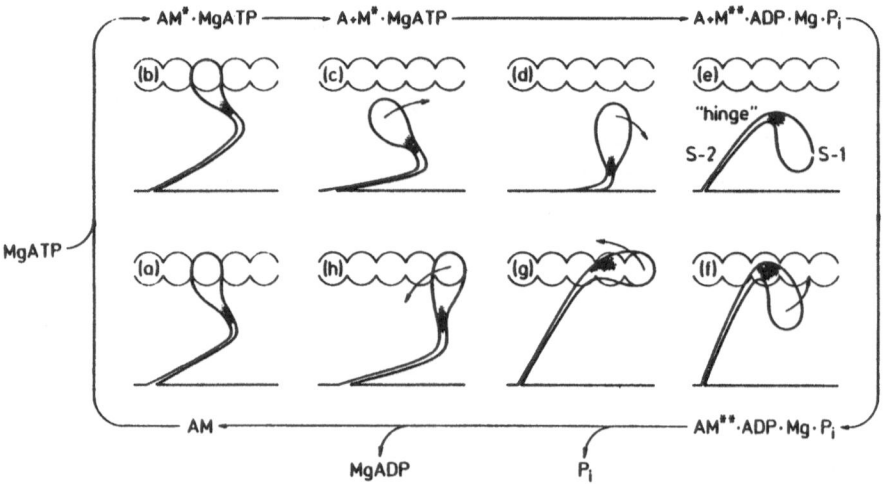

Fig.5.1. A mechanical scheme of ATP hydrolysis during muscle contraction (after [5.3])

between S-1 and S-2. The angle between S-2 and the myosin filament changes passively.[1]

Schemes presented in [5.2,3] have been considered only for the sake of illustration. They are not a bit more or less verisimilar than other concrete assumptions concerning the motion sequences and the relationship between the chemical and mechanical stages of the process (see [5.7] and a detailed review [5.8]). In the preceding chapters it was repeatedly emphasized that the real problems of energy transduction in biological systems, including muscle contraction, are physical problems. To be sure, it would be very interesting to learn the constructions and the exact functioning schemes of many molecular energy transducers in biological systems, but, probably, it is more important to answer first more general question about the principles of their functioning.

5.1.1 The Simplest Quantum-Mechanical Machine Proposed by Gray and Gonda

Perhaps the most important advancement in this field has been the series of works published by GRAY and GONDA [5.9-11]. Although mainly concerned with the functioning of striated muscles, the approaches used are of general character and can be applied to the analysis of any bioenergetic process.

One cannot but agree with the authors' initial statement [5.9] that the physical peculiarity of living systems consists of the fact that these systems contain devices of molecular dimensions capable of performing mechanical work. A biologist,

---

1 It was shown recently [5.5] that X-ray diffraction data can be interpreted in favor of Harrington's model [5.6] according to which the pushing force is generated in the subfragment S-2.

unfortunately, generally associates the word "work" with the word "thermodynamics". At the same time, equilibrium and classical nonequilibrium thermodynamics are completely inadequate when one considers the functioning of machines of molecular dimensions. Analysis of the work performed by a molecular machine requires a quantum-mechanical approach.

Although this problem was comprehensively discussed by FEINMANN more than 40 years ago [5.12], the term "work" can be but seldom found in papers concerning the molecular mechanisms of chemical and biochemical processes (without speaking of rather trivial mention of free energy as a measure of maximal "useful work" that could be performed by a system).

Following GRAY and GONDA [5.9], I shall, therefore, begin with the meaning of forces and work in quantum mechanics. Quantum chemistry as a rule deals with molecular systems whose dimensions are in every case fixed. This means that a fixed system boundary exists, i.e., surfaces, lines or points in space where boundary conditions could be imposed. If boundaries are fixed discrete energy levels and stationary states will, naturally, appear, but no work can be performed until the boundary is shifted. To get work the system boundaries must be moved against the external forces acting on the boundaries.

GRAY and GONDA considered the simplest possible molecular machine of the "particle in a unidimensional well" type. In such a well of width ℓ with impermeable walls the Schrödinger equation in atomic units for a particle with a unit mass (e.g., for an electron) is

$$- \frac{1}{2} \frac{\partial^2 \psi}{\partial x^2} = E\psi \quad , \tag{5.2}$$

with the boundary conditions

$$\psi(0) = \psi(\ell) = 0 \quad . \tag{5.3}$$

The solution of (5.2) consists of eigenvalues and eigenfunctions

$$E_n = \frac{n^2 \pi^2}{2\ell^2} \quad ; \quad \psi_n = \sqrt{\frac{2}{\ell}} \sin \frac{n\pi x}{\ell} \quad ; \quad n = 1,2,\ldots \quad . \tag{5.4}$$

Forces $F_n(0)$ and $F_n(\ell)$ acting on the walls at the points 0 and ℓ are

$$F_n(0) = \frac{\partial E_n}{\partial \ell} = - \frac{n^2 \pi^2}{\ell^3} \tag{5.5}$$

$$F_n(\ell) = - \frac{\partial E_n}{\partial \ell} = \frac{n^2 \pi^2}{\ell^3} \quad . \tag{5.6}$$

The resultant force acting on the system's center of gravity is thus equal to zero, but the walls are subjected to the action of equal and contrarily directed forces (the pressure of electron cloud). Despite these forces the system remains

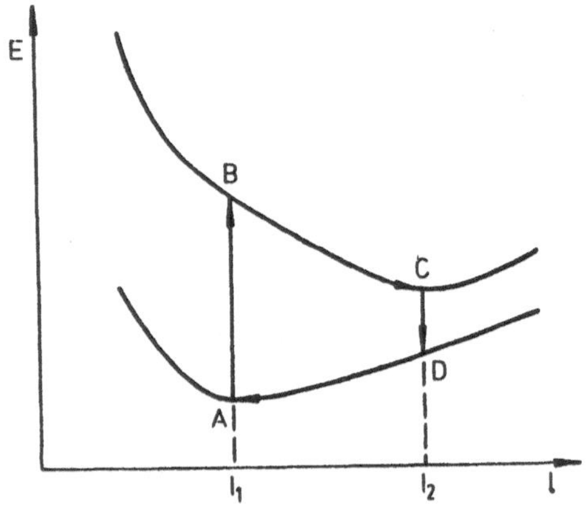

Fig.5.2. The working cycle of the
simplest quantum-mechanical machine

stationary. It means that either the walls have infinite masses or there exist "external" forces acting on the walls and compensating in a stationary state the internal forces. The first assumption is unrealistic, but the second permits us to introduce the notion of work.

Let the electron be initially in a stationary ground state ($n = 1$) in a box of
length ℓ_1. This state corresponds to point A in Fig.5.2. External forces are ba-
lanced by the internal ones which are equal to

$$F_{bal} = \pm \frac{\pi^2}{\ell_1^3} \quad . \tag{5.7}$$

Let the electron be excited from the ground to the excited state with $n = 2$ (point
B in Fig.5.2). The external force remains the same but the internal force instantly
increases, the mechanical balance is upset and the resultant force acting on each
wall becomes

$$F_1 = \frac{3\pi^2}{\ell_1^3} \quad . \tag{5.8}$$

If now at a *proper time moment* (i.e., at the moment of A → B transition) a load
were attached to this system, work could be performed. Let the load correspond to a
force $F < F_1$ (otherwise no work can be obtained) and F = const during the whole working cycle (e.g., the work is the lifting of a constant weight). The displacement of
the walls and the box expansion will lead to internal pressure drop, and a new
mechanical balance will set in at $\ell_2 > \ell_1$:

$$\frac{4\pi^2}{\ell_2^3} = \frac{\pi^2}{\ell_1^3} + F \tag{5.9}$$

(point C in Fig.5.2). The useful work performed along the B → C path equals, obviously,

$$A_{21} = \int_{\ell_1}^{\ell_2} Fd\ell \tag{5.10}$$

and if F = const is, according to (5.9),

$$A_{21} = F(\ell_2 - \ell_1) = \left(\frac{4\pi^2}{\ell_2^3} - \frac{\pi^2}{\ell_1^3}\right)(\ell_2 - \ell_1) \quad . \tag{5.11}$$

This work is performed by utilizing energy ΔE_{21} obtained by excitation:

$$\Delta E_{21} = \frac{3\pi^2}{2\ell_1^2} \quad . \tag{5.12}$$

The efficiency of machine functioning is, naturally, to be estimated from the following expression

$$\eta_{21} = \frac{A_{21}}{\Delta E_{21}} = \frac{2}{3}\, r^2(1 - r)\left(4 - \frac{1}{r^3}\right) \quad , \tag{5.13}$$

where "compression" $r = \ell_1/\ell_2 < 1$.

In a general case r depends on the degree of excitation (i.e., on the quantum number n) and on the load F. The optimum (under given conditions) r value can be determined through η_{21} extremum by equating to zero the η_{21} derivative with respect to r. Using (5.13) we find

$$r_{opt} \approx 0.82 \quad ; \quad \eta_{21}^{(max)} \approx 0.224 \tag{5.14}$$

where $\eta_{n1}^{(max)}$ increases with n and tends to a limit

$$\eta_{\infty 1}^{(max)} = 0.29 \quad . \tag{5.15}$$

The energy (5.12) received by the system is spent on the useful work which, according to (5.11) and taking (5.14) into account, equals $\sim 0.336\ \pi^2/\ell_1^2$, as well as on the work performed against F_{bal}, (5.7). In order to simplify the calculations this force will also be assumed constant. It is easy to show that the work performed against F_{bal} equals $\sim 0.219\ \pi^2/\ell_1^2$.

The system internal energy at point C equals $2\pi^2/\ell_2^2 \approx 1.345\ \pi^2/\ell_1^2$. The sum of these three values is $\sim 1.9\ \pi^2/\ell_1^2$ whereas the system's initial energy at point B prior to the performance of work was $2\pi^2/\ell_1^2$. Where did $\sim 0.1\ \pi^2/\ell_1^2$ units of energy go? At the initial time moment at point B there were large forces acting on both walls: $3\pi^2/\ell_1^3 - F \approx 1.80\pi^2/\ell_1^3$. The wall, therefore, moves with acceleration and acquires kinetic energy. This energy can be assumed to dissipate.

The electron transition at point C back to the ground state (C α D transition on Fig.5.2) leads to a loss of energy $3\pi^2/\ell_2^2 \approx 1.009\pi^2/\ell_1^2$. At the time moment of C → D transition the load must be disconnected, and the last D → A transition is idle. At point D the internal pressure is less than external force F_{bal}, and the D → A transition proceeds under the action of a resultant force, so that a part of the energy received by the system is again transformed into the kinetic energy of the well and dissipates.

The efficiency of this simplest quantum-mechanical machine is not high (~22%). This is caused not by energy dissipation during an irreversible process, but mainly by the large value of energy quantum lost during the C → D transition, i.e., in the final analysis, by the unhappy choice of the system construction (a unidimensional box with one particle, the absence of inner potential field). It is easy to propose a device (e.g., a three-dimensional box with one movable wall), for which points C and D lie close enough to one another (BC and AD can even intersect), and the efficiency may reach 80%-90%.

In the course of our discussion of this model three assumptions have been tacitly made.

1) Excited state B lives long enough to have time for the B → C transition to be completed. In other words, the B → A transition can be neglected. It is clear, therefore, that the A → B transition cannot be regarded as a usual electron or vibration excitation of the system. The characteristic time for a myosin bridge during muscle contraction (i.e., characteristic time of the working stroke B → C in our model) is ~10^{-3} s. The life times of the usual electron and vibration excitations are considerably less. The first question that arises when one discusses the functioning of any quantum-mechanical machine is that of the physical nature of the primary excited state. It is easy to see that this question in essence refers to the physical nature of the long-living conformationally nonequilibrium state arising in a macromolecular system after a sufficiently fast perturbation (Chap.4).

2) Automatic attachment of the load after a fast A → B transition and its disconnection after C → D transition were postulated. The mechanism of these regulatory stages without which the machine functioning is impossible, in other words the control mechanism, will be discussed below (Sect.5.1.3).

3) The adiabaticity of a quantum-mechanical process was assumed, i.e., the possibility of the system subdividing into two subsystems, a fast and a slow one, so that the local equilibrium of the fast subsystem is reached at fixed parameters of the slow subsystem. This is in essence the problem of the Born-Oppenheimer approximation in which the validity of adiabatic treatment is justified by the large difference between the masses of nuclei and electrons. GRAY and GONDA hold that in the case of their model the validity of adiabatic approximation follows from the comparison between the characteristic times of the working stroke (~10^{-3} s) and of electron excitation (Bohr's period of a quantum corresponding to ATP hydrolysis free energy equals ~10^{-14} s). This comparison with an essentially thermodynamic characteristic of a process, free energy, seems, from my point of view, to be not fully consistent. The physical nature of the fast and slow subsystems, as well as of the role played

100

by the system's thermodynamic free energy in the functioning of an individual molecular device, deserve special discussion (Sect.5.3.2,3). Let us consider now a somewhat more complicated molecular machine also proposed by GRAY and GONDA [5.9].

5.1.2 Model: Two Particles in a Box with a Movable Partition

This model is shown in Fig.5.3 [5.9]. For simplicity both compartments (a and b) are assumed to be initially identical. In state A both particles occupy the lowest levels, the system is at equilibrium, the pressures on the partition from the left and from the right are equal. Fast excitation of the particle in (a) (A → B transition) transfers the system into a nonequilibrium state, the pressure from the left is now stronger than from the right, the partition is shifted (B → C transition), and if a proper load is attached, work can be performed. The partition moves to the right until the pressures are again equilibrated. A part of the energy is stored now in compartment (b), which is never excited during the cycle and plays the role of a flywheel returning the system into its initial state after fast de-excitation in compartment (a) (C → D and D → A transitions). During the D → A transition work can be again performed. It is not difficult to make calculations similar to those in the preceding section.

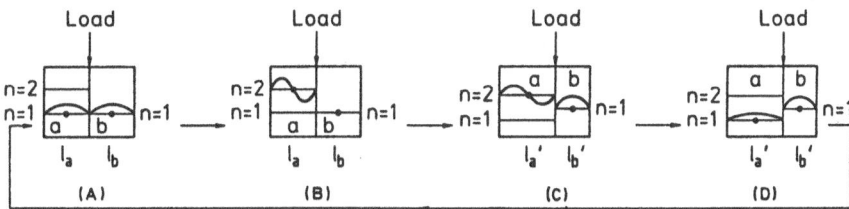

Fig.5.3. Model: Two particles in a box with a movable partition

5.1.3 Automatic Regulation in a Model of the Quantum-Mechanical Machine

The most significant contribution made by GRAY and GONDA [5.9,10,13] is, probably, their analysis of automatic regulation and control during the functioning of molecular machines. According to GRAY and GONDA, there are two principal questions here.

1) The probability of transition between two states A and B under the action of perturbation V equals in quantum mechanics $(\int \psi_A \bar{V} \psi_B d\mathscr{T})^2$ and is a symmetric function relative to the A and B states. Is it possible to neglect the backward transition, B → A, in the model shown in Figs.5.2,3?

2) How can one ensure the well-timed attachment and detachment of a load so that there is no useless energy dissipation?

Following [5.9], let us begin with the second question. We have already seen in Sect.4.1 that the solution of the control problem quite satisfactory for macroscopic machines (a thermodynamically reversible process) does not work in the case of molecular machines [5.14]. None of the numerous schemes of muscle contraction proposed

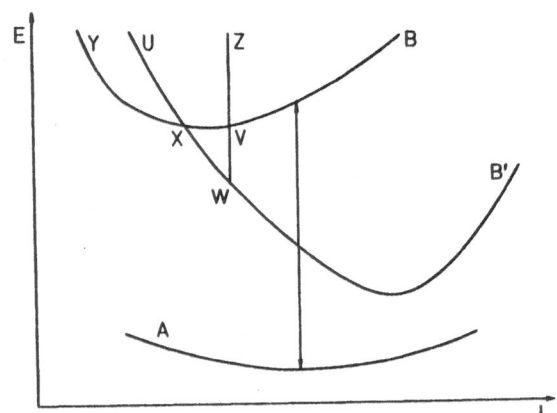

before Gray and Gonda even tried to formulate the problem of a control physical mechanism (except, perhaps, very general reasonings in [5.15,16]).

Quantum-mechanical analysis carried out in [5.9] shows how synchronization can be ensured between excitation and load without direct control. This analysis is based on the quantum-mechanical concept of induced transition [5.13]. Let us assume that two excited states, B and B' exist (Fig.5.4). The A → B excitation is reversible, but in state B the system cannot perform work. The "working" excited state B' intersects B at point X (in a real molecular machine, as distinct from our simplified model, potential curves are to be replaced by surfaces in multidimensional space). A force exerted by the machine at a given coordinate value ℓ is proportional to the slope of the B' curve at this point. The machine excitation into a "pushing" state proceeds thus in the neighborhood of X where the energies of states B and B' are close, so that a radiationless transition is possible. The theory of these transitions is well known. The probability of radiationless transition between B and B' states can be estimated with the help of the following formula:

$$P_{B \to B'} = \frac{2\pi}{\hbar} C^2 S^2 p \qquad (5.16)$$

where C is the electron coupling factor of states B and B' whose value is solely determined by the nature of these states and can be considered as fixed; here S is the so-called Frank-Condon overlap integral between two vibrational wave functions in states B and B'; and p is the density of vibration levels in the B' electron state. The S factor can depend significantly on the load and, therefore, the role of perturbation which transfers the system into a working state can be played by the load itself. One of the possible ways to ensure this is described in [5.9]. The oscillator path in state B is barred by an obstacle — a potential barrier that forces the system to spend more time in the neighborhood of the intersection point X and thus increases the amplitude of vibrational wave function in this region.

In Fig.5.4 this barrier is represented as a change of the potential curves (YXVZ instead of YXVB for state B and UXWZ instead of UXWB' for state B'). The overlap integral S increases and the probability of machine transition to the working state $P_{B \to B'}$ rises. It is thus unnecessary to attach a load in a controllable way at a fixed time moment, as the stochastically realized attachment of a load automatically starts the working cycle.

Scheme 5.4 has a clear physical meaning. Coordinate ℓ determines the position of a force-generating particle: of the electron in a box in the model of the simplest quantum-mechanical machine (Sect.5.1.1), of the head of a myosin bridge hinge in the model of a striated muscle.

The coordinate of a barrier ZW represents the position of the movable partition in the quantum-mechanical model, or the position of the fixed actin monomer in the muscle model. The barrier moves slowly compared with the motion of the particle, and its position, which determines the shape of the system's potential curve, can be re-garded as a parameter. This makes it possible to solve the equation of motion e.g., (5.2) for the simplest model at various fixed values of this parameter. It is a standard procedure in all cases when a system can be naturally subdivided into fast and slow subsystems. Such subdivision is justified if the "barrier" is much heavier that the "particle". It should be noted that the term "heavier" does not necessarily relate to the comparison of the true masses of "barrier" and "particle". The "barrier" effective mass may be large and its motion slow, due to the fact that this motion represents a conformational transition in a complex macromolecular sys-tem and is hindered by a high entropy barrier.

Analysis shows that a relatively slow "barrier" motion means that for every adi-abatic state of a "particle" there exists a multitude of practically continuous energy levels of the "barrier". This fact leads to the practical irreversibility of the B → B' transition at point X in Fig.5.4. As a matter of fact, it is the transi-tion from *one* level B to one of many B' levels.

The principle of microscopic reversibility is, naturally, valid. The probability of the transition of a system from B to a *given* B' level is equal to the probability of back transition. In the case of B-B' transition, however, it does not matter to which one of the many B' levels a system will be transferred. The probability of transition to a given B' level is very low (there are many almost identical levels), but the probability of transition to *any* of the B' levels is quite high. The probabi-lity of backward transitions to the initial B level is not zero, and sometime the system would have to return to the initial state. This "sometime", however, for a more or less complex molecular system, may not come for a very long time. A transi-tion to any B' level leads to a rise in the pushing force, the "barrier" begins to move slowly along the B' curve, work is performed, and the system's energy has enough time to decrease so much that the backward transition becomes practically impossible.

GRAY and GONDA [5.10] proposed a concrete model of a chemomechanical energy trans-ducing device in striated muscle of vertebrates, based on the physical principles formulated above. In this model the role of transducer is played by the myosin bridge.

The excitation of the machine [A → B transition in schemes (5.2,4)] represents the reversible ATP attachment to the catalytic center of the myosin bridge, and the irreversible transition to the "working" curve [at point X in scheme (5.4)] is induced by the attachment of a load, i.e., by the bridge attachment to actin filament. The working stroke of the machine is realized as a slow thermodynamically irreversible conformational change of the myosin bridge, and the breakaway of reaction products (ADP and P_i) is in accordance with the usual mechanism of fast reversible dissociation. The second radiationless transition — the myosin bridge detachment and the return of the machine to its initial state — is also comprehensively discussed. I think that this work solves the main *physical* problems of muscle contraction. Subsequent investigation may of course amend certain details of the relationship between the biochemical and the mechanical events, and introduce certain changes into the concrete construction. I cannot present here the whole article [5.10], it would have to be re-copied. I highly recommend the reader to study this paper: it will be a true aesthetic pleasure.

This analysis thus shows that the functioning of an elementary chemomechanical transducer in the course of muscle contraction can be described in the realm of relaxation concept: fast local perturbation leading to a rise in an essentially nonequilibrium conformation state, and slow directed relaxation to the new equilibrium in the course of which useful work is performed.

5.2 Active Transfer of Ions[2]

The energy aspects of active ion transfer were discussed in Chap.2 with the membrane sodium pump taken as an example. We have seen that the transfer of Na^+ and K^+ ions through a cell membrane is realized with the help of a special molecular machine — the Na-, K-ATPase. ATP hydrolysis serves as an energy-donating process, and the stoichiometry of an elementary act as a rule corresponds to the transfer of three Na^+ ions from the cell and two K^+ ions into the cell per one ATP molecule hydrolyzed.

There are various concrete models of Na-, K-ATPase described in literature [5.17-19]. The true construction of this molecular machine is not known. One of the recently proposed models, which is in good agreement with the known experimental facts, will be described below [5.20-23]. As in all molecular machines the energy-donating reaction not only ensures the energy-consuming process but itself depends essentially on it, i.e., on the load attachment. In the absence of any of the transferred ions (Na^+ or K^+) ATPase activity practically vanishes. It was shown that the enzyme activation by Na^+ ions is bound up with polar hydrophilic regions of a macromolecule, whereas the activation by K^+ ions is connected with hydrophobic regions. The Na-, K-ATPase model proposed in [5.21] is shown in Fig.5.5. Strictly speaking

---

2 This section has been written together with V.A. Tverdislov and L.V. Yakovenko.

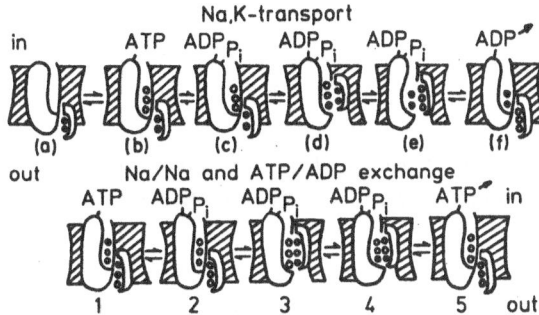

Na,K-transport

Fig.5.5. Model of Na-, K-ATPase (after [5.21])

this figure represents only one dimer with one large and one small subunit. A real enzyme contains at least two such dimers.

It is assumed that in the absence of ATP each subunit binds cations independently. Cations fill special ion-exchange cavities that can hold 3 Na^+ ions or 2 K^+ ions. The selectivity of small subunit is not essential — ion binding can be completely nonspecific. The total charge of this ion-exchange cavity makes the binding of three cations more favorable than the binding of two cations.

The ion-binding specificity of the large subunit changes during the reaction cycle and is determined by the enzyme conformational state. Let us begin with the functioning of ATPase in the regime of Na/K exchange. This means that the small subunit is assumed to contain initially 2K^+ ions. In the absence of ATP, the large subunit will bind predominantly neither K^+ nor Na^+ ions. Under these conditions ion binding is of the equilibrium type and is determined by the association constants. This state of the enzyme is shown in Fig.5.5a.

In the presence of ATP enzyme phosphorylation becomes possible; the formation of a stable phosphorylated intermediate requires, however, the binding of 3 Na^+ ions by the large subunit. The binding of 3 Na^+ ions is facilitated by the conformational changes after ATP attachment, leading to an increase in the large subunit affinity to the Na^+ ions. ATP attachment also leads to an increase of the ion-exchange cavity effective charge caused, e.g., by the negatively charged γ-phosphate group of ATP. After ion exchange with intracellular medium the large subunit cavity becomes occupied by three Na^+ ions (Fig.5.5b), and phosphorylation of aspartic acid residue at the catalytic center of the large subunit takes place. As a result of subsequent conformational relaxation, three Na^+ ions become locked within the ion-exchange cavity of the large subunit. These structural rearrangements lessen once more the large subunit affinity to ions. State 5.5b becomes energetically unfavorable but kinetically stabilised. The relaxation of this state leads to the formation of a stable phosphorylated enzyme (Fig.5.5c). The b → c stage is practically irreversible: the backward reaction requires the overcoming of a high activation barrier.

Subsequent steps of reaction involve the small subunit. The dimer subunits are assumed to be able to shift slightly with respect to each other so that the ion-exchange cavity of the small subunit communicates either with the outside medium or with the ion-exchange cavity of the large subunit. In the latter case ion exchange between subunits becomes possible. In the realm of this model, the subunit is displaced not under the action of a directed force, but as a result of stochastic thermal fluctuations in the direction determined by enzyme construction.

During the lifetime of the stable phosphorylated state the situation favorable for ion exchange between subunits is repeatedly realized. If in any of these cases K^+ ions happen to be in the cavity of the small subunit there appears a state shown in Fig.5.5d. In this state, ion exchange between subunits is energetically favorable. The after-exchange state is shown in Fig.5.5e. The $d \rightleftharpoons e$ equilibrium is shifted to the right owing to the high affinity of the large subunit to K^+ ions. The exchange of 3 Na^+ ions for 2 K^+ ions in the large subunit leads to a rise in a conformationally nonequilibrium state of the whole system. Conformational relaxation begins. At the first stages of relaxation, the stochastic ion exchange between subunits is still possible, but, beginning with a certain degree of conformational changes, it is terminated because of steric hindrances, and the $d \rightarrow e$ process becomes irreversible. A decrease in the positive charge within the large subunit cavity is accompanied by the disappearance of a high potential barrier for enzyme dephosphorylation (dissociation of P_i anion). The large subunit cavity opens into intracellular space (Fig.5.5f). After ADP dissociation the enzyme returns to its initial state.

According to this model, various regimes of Na-pump functioning are possible: Na/K, Na/Na, K/K exchange, passive ion transfer, reversed regime of Na-, K-ATPase, ion transfer with different stoichiometry. All these regimes contribute to the overall process of ion transport. For any of them to be realized in a "pure form" specific requirements must be met. In the absence of K^+ ions on both membrane sides the regime of Na/Na exchange is realized. Ion exchange between the subunits (3 Na^+/3 Na^+) does not lead in this case to subsequent relaxation. As distinct from Na/Na exchange, the coupled Na/K exchange proceeds far away from equilibrium so that $b \rightarrow c$ and $d \rightarrow e$ stages are kinetically irreversible.

We have seen in Sect.2.2 that a reversed ATPase reaction, i.e., ATP synthesis from ADP and P_i, can take place without the formation of concentration gradients of these ions with reversed signs. A fast increase in Na^+ ion concentration (a "sodium jump") leads to the appearance of a nonequilibrium state: the large subunit, whose conformation corresponds to the absence of Na^+ ions in the cavity, binds these ions fast enough. Relaxation of this state ensures ATP synthesis. It should be borne in mind that during this process the system does not pass the same intermediary in backward order, as happens during the $b \rightarrow c$ stage in the course of Na/K and Na/Na exchange.

5.3 Enzymatic Catalysis

5.3.1 Relaxation Concept of a Catalytic Act

Enzymatic reactions are obligatory participants of all bioenergetic processes. A minute analysis of enzyme functioning in the realm of relaxation concept (as well as of other theoretical concept) can be found in [5.24]. Certain physical aspects of this problem will be discussed here, and using an enzymatic reaction as an example, the important question of the role played by the free energy of macroscopic systems in the functioning of molecular machines will be considered.

The relaxation concept of enzymatic catalysis was formulated in 1972 [5.25,26]. According to this concept conformational relaxation of the substrate-enzyme complex is, essentially, an elementary act of enzymatic reaction, and the rate of substrate-product transformation is determined by the rate of this relaxation. Conformational relaxation of the substrate-enzyme complex, initiated by the substrate attachment to the enzyme active center, includes not only the rupture of the old and formation of new secondary bonds in the protein globule, but also the chemical changes necessary to transform the substrate molecule into a molecule or molecules of a product.

The principal idea of the relaxation concept must be underlined. It is not simply a question of the substrate-enzyme complex relaxation, in the course of which the enzyme reactivity changes. A scheme of this type has been proposed in the pioneer work by SIDORENKO and DESHCHEREVSKY [5.27] and a more strict quantitative analysis in [5.28]. The main idea of the relaxation concept, as it was formulated in [5.25], is that conformational relaxation initiated by the substrate binding proceeds under the action of a force that pushes the chemical system directionally along the reaction coordinate. An act of the substrate-product chemical transformation is a compulsory constituent of the conformational change of the substrate-enzyme complex.

A simplified general scheme of an enzymatic reaction can be written as follows

$$
\underset{\substack{-S \\ a \\ (fast)}}{\overset{\substack{+S}}{E \rightleftharpoons}} \underset{\substack{b \\ (slow)}}{SE} \rightarrow \underset{\substack{c \\ (fast)}}{\overset{\substack{-P}}{P\tilde{E} \rightleftharpoons}} \underset{\substack{+P}}{\tilde{E}} \underset{\substack{d \\ (slow)}}{\longrightarrow} E \quad . \tag{5.17}
$$

Fast acts of reversible stages (a) and (c) are realized at an almost unchanged enzyme conformation. The overall process rate is determined by slow irreversible relaxation stages (b) and (d) (conformational relaxation of the substrate-enzyme complex and of the free enzyme, respectively). The change of the system in the course of these stages may be described as mechanical motion along a specific degree of freedom, brought to a stop after a new equilibrium state is reached. This mechanism, naturally, is only effective far away from the thermodynamic equilibrium of the chemical system S-P. The motion along a specific degree of freedom proceeds under the action of a force arising between the relaxed and nonrelaxed regions of the macromolecular complex after perturbation of the enzyme active center by a substrate molecule (Chap.4).

5.3.2 Coherent Phonons and Catalytic Transformation

The relaxation concept was in a sense a starting point for an endeavour to construct a physical theory of enzymatic catalysis in an extremely interesting work by FAIN [5.29], which, I think, has been an essential advance towards a quantitative theory of biocatalytic processes. The main content of [5.29] will be qualitatively set forth below.

The author states that the usual assumption underlying all the traditional approaches to the kinetics of chemical reactions, according to which all the vibrational degrees of freedom undergo fast relaxation to thermodynamic equilibrium, do not hold true under certain conditions. In the case of ordered macromolecular structures, surplus electron energy may be transformed in the course of radiationless relaxation not into heat but into the energy of coherent vibrational motion. At first sight the situation resembles that in quantum generators, where a transition between levels is accompainied by the excitation of coherent electromagnetic oscillations. As in the case of quantum generators it is necessary to maintain an inverse population of levels during the process: the population of the higher energy level must be greater than the population of the lower one. The main difference between an enzymatic system and a quantum generator lies in the fact that in the latter the frequency of excited electromagnetic oscillations approximately coincides (to within the linewidth) with the frequency of electron transition. In the case of a substrate-enzyme complex, the inverse population is maintained owing to continuous substrate influx.

The oscillations excited are not electron oscillations, but the vibrations of a nuclear subsystem coupled with the electron subsystem by electron-phonon interaction. The electron transition frequency is in this case much higher than the frequency of any vibration of the nuclear subsystem. It means that excitation of nuclear vibrations may occur only through a multitude of quantum transitions. In the case of a quantum generator this is impossible because of the smallness of electron-photon interactions, but electron-phonon interactions are not small, and the transition between the remote states of a system are realized by the excitation of coherent vibrations of the nuclear subsystem (in other words, through conformational changes of a macromolecule). Transition between the end electron states of the system during catalytic transformation [using the language of scheme (5.17), the transition between states S + E and $P\tilde{E}$] is realized by the excitation of coherent vibrations as a result of substrate attachment. The probability of their thermal excitation is negligibly low.

The main theoretical achievement of this work is probably the selfconsistent description of the simultaneous changes taking place in the electron and nuclear subsystems. In all preceding theories of chemical kinetics the changes in electron states and in nuclear vibration amplitudes were described separately.

Equations obtained in [5.29] for the reaction rate dependence on substrate concentration S lead to the conclusion that at low enough S values the reaction rate abruptly becomes negligibly low. The sharp (not in accordance with the mass action law) drop of reaction rate with S decrease is one of the important predictions made in [5.29]. Another theoretical prediction that can be checked experimentally is the possibility to observe coherent conformational oscillations of the substrate-enzyme complex. Both these predictions could, evidently, be made by qualitative consideration of schemes like (5.17) [5.24,28].

In the realm of classical chemical kinetics the basic idea of [5.29] (as well as of the relaxation concept in general) can be formulated as follows. A substrate-product transformation requires the overcoming of a certain activation barrier. The excitation of coherent nuclear vibration, i.e., the excitation of the specific degree of freedom coinciding with the reaction coordinate (here, as a matter of fact, lies the enzyme specificity) is equivalent to the creation of the inverse population at the system's higher and lower levels. In other words, the specific degree of freedom has a very high local temperature. The probability of its being "changed" for noncoherent vibrations of the nuclear subsystem is negligibly small, and the most probable relaxation path includes the chemical substrate-product transformation. Chemical transformation is thus realized at a very high local temperature, the effective kT value is much higher than the reaction activation barrier, which under these conditions cannot be regarded as an obstacle.

It must be stressed that the energy increase, "heating", concerns only the specific mechanical degree of freedom. Only the energy localized at this degree of freedom is increased in a conformationally nonequilibrium state and lowered in the course of conformational relaxation. The change in the total energy of the substrate-enzyme system caused by the complex formation may have any sign. What is important is that owing to substrate binding there appears one (but not one of the many possible) nonequilibrium state with local steric constraint, a specific mechanical degree of freedom is excited.

Another approach is also possible. If we do not single out the mechanical degree of freedom but consider one macromolecular complex SE in (5.17) as a statistical system [5.30], then the most important occurrence at stage (b) is the entropy increase: the SE state relative to the final $\widetilde{PE}$ state is an extremely improbable fluctuation.

5. 3. 3 Free Energy Change During Chemical Transformation and the Conditions in which Machine Mechanisms are Advantageous

As already mentioned repeatedly, machine mechanisms in bioenergetics can be realized only far from the state of thermodynamic equilibrium. As the chemical equilibrium of an S-P system is approached the rate of relaxation stages in (5.17) sharply decreases. The source of energy required for the origin and maintenance of a force that pushes a substrate-enzyme complex along the reaction coordinate is, in the final analysis, the difference between the free energy values of the initial and the final reaction products. It is owing to this very energy that the excitation of a specific degree of freedom is realized. Naturally if a decrease in the free energy of a chemical system involves its entropy, the concentration component, then the energy required to carry out every elementary act is derived from the thermostat (Sect.3.2). At thermodynamic equilibrium mechanical degrees of freedom cannot be excited. Mechanical motion comes to a stop: there is no dynamic equilibrium in mechanics (Sect.4.3).

Let us try now to answer the following question: why and under what conditions is a machine good? In other words, in what cases is the relaxation mechanism capable of ensuring a higher efficiency of the catalytic process than that obtained in conventional Arrhenius-type mechanisms where the activation barrier is overcome in every elementary act owing to thermal fluctuations? In the case of a perfect enzymatic machine, the free energy change caused by the substrate-product transformation is completely utilized to excite a specific degree of freedom coinciding with the reaction coordinate. In the case of Arrhenius' purely statistical mechanism, this degree of freedom (as well as any other) will get RT/2 of thermal energy (per one mole). A temperature at which the efficiency of Arrhenius' mechanism would be equal to that of a machine, for an ideal system, can be found using equation

$$T = - \frac{2\Delta G}{R} \quad . \tag{5.18}$$

The ΔG values for the majority of intracellular enzymatic reactions are usually in the neighborhood of -10 kcal/mole. If we accept this value for ΔG, then $T \approx 10^4$ K. It means that even with a 20% efficiency of a molecular machine (and the efficiency may be much higher, see Sect.5.1.1) a machine-like mechanism ensures the reaction rate, that with Arrhenius' mechanism could be reached at $T \approx 2000$ K (provided the system is able to function at such a high temperature).

We can conclude that machine mechanisms are especially favorable in the cases of energy coupling of reactions *every one of which* is accompanied by a considerable change in free energy, or in the cases of enzymatic reactions with high G values for substrate-product transformation. It is for these very systems that the machine mechanisms are, most probably, realized.

5.4 Membrane Phosphorylation

Analysis of the existing experimental data carried out in the preceding chapters shows beyond any doubt that light-dependent, as well as dark, membrane phosphorylation processes are realized by means of molecular machines functioning far away from equilibrium. The working stroke of such a machine, as an elementary act of the energy transducing process, must be the thermodynamically irreversible relaxation of a macromolecular structure that can be described as mechanical motion. In the case of membrane phosphorylation the energy-donating reaction is redox transformation of an ETC carrier-transformer (electron transfer through a coupling site), and the energy consuming reaction is ATP synthesis from ADP and P_i catalyzed with ATP-synthetase.

A possible model of the molecular machine able to ensure this coupling will be described below. We use here the model of a unidimensional box with a movable partition which has been proposed by Gray and Gonda for the muscle contraction process (Sect.5.1.2).

5.4.1 Model of a Redox Molecular Machine

The model proposed by Gray and Gonda, as well as all the necessary initial formulas, have already been considered in Sects.5.1.1,2. In our case, the model represents a carrier-transformer whose conformational relaxation is accompanied by considerable energy decrease [5.31].

The ETC electron transfer from a donor D to an acceptor A through a carrier-transformer T can be represented by the following scheme:

$$D_1^- T_2 A_2 \xrightarrow{fast} D_1 T_2^- A_2 \xrightarrow{slow} D_2 T_1^- A_2 \xrightarrow{fast} D_2 T_1 A_2^- \xrightarrow{slow} D_2 T_2 A_1^- \quad . \tag{5.19}$$

Here superscript "minus" designates a reduced state of the active center, and subscripts "1" and "2" refer to the remaining part of the protein globule in the relaxed (equilibrated) reduced and oxidized states of a given carrier, respectively. The fast stages $a \to b$ and $c \to d$ require a rather precise coincidence of the corresponding occupied and free electron levels of neighboring carriers [5.32]. Useful work can be performed in the course of slow relaxation stages $b \to c$ and $d \to e$.

A scheme of the changes in the states of a carrier-transformer during electron transfer through a coupling site is shown in Fig.5.6. Box A represents the carrier active center undergoing redox transformations, and box B the remaining protein globule. For simplicity the initial lengths of the boxes, ℓ_0, are assumed to be equal, and the electron systems of each box are represented by an electron pair on the lowest orbital.

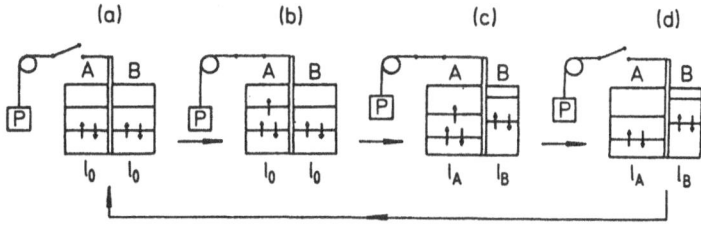

Fig.5.6. A model of the redox molecular machine

In the initial state a the system is at equilibrium, energies $E_A^{(a)}$ and $E_B^{(a)}$ of subsystems A and B, and overall energy $E^{(a)}$ are

$$E_A^{(a)} = E_B^{(a)} = \frac{\pi^2}{\ell_0^2} \quad ; \quad E^{(a)} = 2 \frac{\pi^2}{\ell_0^2} \quad . \tag{5.20}$$

The forces acting on the partition from the left and from the right are, respectively

$$F_A^{(a)} = F_B^{(a)} = 2\,\frac{\pi^2}{\ell_0^3} \quad . \tag{5.21}$$

Let the carrier-transformer active center now be reduced with an electron transferred from the preceding ETC carrier to the lowest free orbital of subsystem A (state "b"). The energy of subsystem A increases and the energy of subsystem B does not change:

$$E_A^{(b)} = 3\,\frac{\pi^2}{\ell_0^2} \quad ; \quad E_B^{(b)} = \frac{\pi^2}{\ell_0^2} \quad ; \quad E^{(b)} = 4\,\frac{\pi^2}{\ell_0^2} \quad . \tag{5.22}$$

State "b" is conformationally out-of-equilibrium: the forces acting on the partition from the left $F_A^{(b)}$, and from the right, $F_B^{(b)}$, are not equal to each other:

$$F_A^{(b)} = 6\,\frac{\pi^2}{\ell_0^3} \quad ; \quad F_B^{(b)} = 2\,\frac{\pi^2}{\ell_0^3} \quad . \tag{5.23}$$

A conformational relaxation begins: the partition will be shifted to the right under the action of a variable resultant force that was initially equal to

$$F_1 = 6\,\frac{\pi^2}{\ell_0^3} - 2\,\frac{\pi^2}{\ell_0^3} = 4\,\frac{\pi^2}{\ell_0^3} \quad . \tag{5.24}$$

In the course of conformational relaxation useful work can be performed, e.g., the lifting of a weight, the drawing together of electric charges of the same sign, etc. For this purpose at an appropriate moment of time (best of all at the moment of a → b transition) a load should be attached. The problem of automatic load attachment and disconnection has been discussed in Sect.5.1.3.

Let the load correspond to a constant force $F < F_1$. The partition shift leads to the new equilibrium state "c", in which

$$F_A^{(c)} = 6\,\frac{\pi^2}{\ell_A^3} = F_B^{(c)} = 2\,\frac{\pi^2}{\ell_B^3} + F \quad . \tag{5.25}$$

The useful work performed is

$$A_{bc} = \int_0^{(\ell_A - \ell_B)/2} F\,dx \quad , \tag{5.26}$$

where x is the partition displacement from its initial position at a given moment of conformational relaxation. Substituting F from (5.25) into (5.26) we obtain

$$A_{bc} = \pi^2 \left(\frac{3}{\ell_A^3} - \frac{1}{\ell_B^3} \right) (\ell_A - \ell_B) \quad . \tag{5.27}$$

This work is performed by utilizing the energy ΔE_{ab} received by subsystem A as a result of the a → b process:

$$\Delta E_{ab} = E^{(b)} - E^{(a)} = 2 \frac{\pi^2}{\ell_0^3} \quad . \tag{5.28}$$

The fraction of this energy transformed into work is

$$\eta_{bc} = \frac{A_{bc}}{\Delta E_{bc}} = \frac{1}{8}\left(4 - r^3 + \frac{3}{r} - r^2 - \frac{3}{r^2} + r - \frac{3}{r^3}\right) \quad , \tag{5.29}$$

where

$$r = \ell_A/\ell_B \quad . \tag{5.30}$$

Solution of equation $\partial \eta_{bc}/\partial r = 0$ yields r_{opt} for this (rather poor) machine construction:

$$r_{opt} \approx 1.20 \quad ; \quad \eta_{bc} \approx 0.09 \quad ; \quad \ell_A \approx 1.09\, \ell_0 \quad ; \quad \ell_B \approx 0.91\, \ell_0 \quad . \tag{5.31}$$

The useful work performed is

$$A_{bc} \approx 0.18 \frac{\pi^2}{\ell_0^2} \quad . \tag{5.32}$$

The total energy of the system in the state "c" is

$$E^{(c)} = 3\frac{\pi^2}{\ell_A^2} + \frac{\pi^2}{\ell_B^2} \approx 3.72 \frac{\pi^2}{\ell_0^2} \quad . \tag{5.33}$$

In the course of relaxation b → c the system inner energy has thus decreased by $0.28\, \pi^2/\ell_0^2$. About $0.18\, \pi^2/\ell_0^2$ was spent for useful work. The remaining energy was transformed into partition kinetic energy and, in the end, dissipated. Owing to b → c relaxation the energy of an electron on the highest occupied orbital of subsystem A decreased by

$$\Delta E_{el} = 2\pi^2\left(\frac{1}{\ell_0^2} - \frac{1}{\ell_A^2}\right) \approx 0.32 \frac{\pi^2}{\ell_0^2} \quad . \tag{5.34}$$

The energy of an electron on the highest occupied orbital of subsystem A now coincides approximately with the energy of a free orbital of the next ETC carrier, and the electron can leave the carrier-transformer (stage c → d).

The total inner energy in state "d" is

$$E^{(d)} = \frac{\pi^2}{\ell_A^2} + \frac{\pi^2}{\ell_B^2} \approx 2.04 \frac{\pi^2}{\ell_0^2} \quad . \tag{5.35}$$

This state is again out-of-equilibrium. The forces acting on the partition from the left $(F_A^{(d)})$ and from the right $(F_B^{(d)})$ are

$$F_A^{(d)} = \frac{\pi^2}{\ell_A^3} \approx 1.54 \frac{\pi^2}{\ell_0^3} \quad ; \quad F_B^{(d)} = 2\frac{\pi^2}{\ell_B^3} \approx 2.67 \frac{\pi^2}{\ell_0^3} \quad . \tag{5.36}$$

Let the load be disconnected at the time moment of c → d transition (Sect.5.1.3). The second relaxation stage d → a will, therefore, not be accompanied by useful work. The energy decrease

$$\Delta E_{ad} \approx 0.04 \frac{\pi^2}{\ell_0^2} \tag{5.37}$$

will be transformed into the partition kinetic energy and be dissipated.

The energy of an electron has thus been decreased in the course of its passage through the carrier-transformer by ~0.32 π^2/ℓ_0^2. About 0.18 π^2/ℓ_0^2 of this energy was spent on useful work at relaxation stage b → c, and the remaining energy (~0.14 π^2/ℓ_0^2) dissipated in the course of two relaxation stages, b → c and d → a. The resultant efficiency of redox transformation energy being transduced into useful work is ~56%.

The above calculations, naturally, have an illustrative character. The numerical values obtained depend essentially on the model construction, i.e., on the shape and dimensions of a potential box in Fig.5.6.

5. 4. 2 Certain Corollaries Arising from the Relaxation Concept of Membrane Phosphorylation

If an energy-transducing act of membrane phosphorylation actually coincides with the act of a conformational relaxation of carrier-transformer, then the possibility to observe the conformationally nonequilibrium states should depend on the conditions in which the electron transport proceeds. Let us return to the schemes in Figs.5.5 and 5.6. At the beginning of conformational relaxation b → c, the carrier-transformer is in a reduced nonequilibrium state "b", and at the end of relaxation in a reduced equilibrium state "c". At all other moments of time the carrier-transformer is oxidized. Let us assume that we register the carrier-transformer in its reduced state. The probability to find it in a conformationally nonequilibrium state depends, in this case, on the relationship between the rate of conformational relaxation b → c and the frequency of electron transfer acts c → d. When the limiting stage of the overall process is conformational relaxation (i.e., according to the relaxation concept, the phosphorylation reaction) the probability to find the carrier-transformer in an equilibrium state is low: most of the time the carrier exists in a not completely relaxed state and transfers the electron further along the ETC immediately after relaxation. On the other hand, when the limiting acts are those of electron transfer, the reduced carrier-transformer exists most of the time in equilibrium state "c".

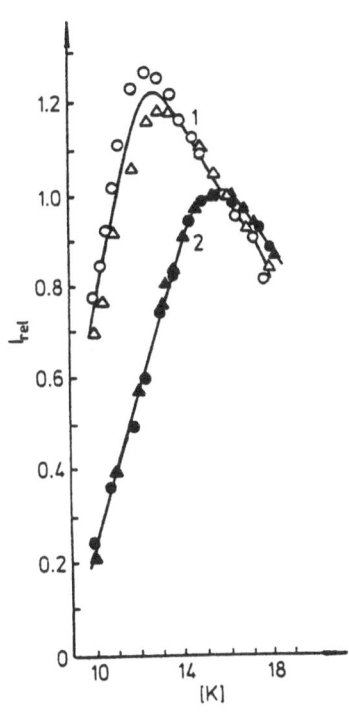

Fig.5.7. I(T) curves of the EPR signal of N-2 center isolated in beef heart mitochondria. (1)o Uncoupled mitochondria reduced with thermalized electrons at 77 K; (2)● the same preparation after relaxation at 210 K; (1)△ active phosphorylating mitochondria in the state of respiratory control; (2)▲ the same preparation after uncoupling by the freezing and thawing procedure (after [5.31])

The clearest confirmation of the aforesaid was obtained in a study of the mitochondrial first coupling site [5.31,33,34]. The carrier-transformer is iron-sulfur center N-2 (Sect.2.4) that gives an EPR signal in its reduced state. In Sect.4.2 it was shown that a reduced N-2 center and other ferredoxin-like proteins as well are characterized by different spin-lattice relaxation times in their equilibrium and nonequilibrium states, which leads to differences in the temperature dependences I(T)-curves of their EPR signal intensities at a fixed microwave power.

Figure 5.7 shows I(T) curves of the reduced N-2 center in beef heart mitochondria taken under different conditions. After reduction of the carriers in the frozen oxidized uncoupled mitochondria with thermolized electrons at 77 K, a conformationally nonequilibrium reduced N-2 center appears (curve 1). Subsequent incubation at 210 K leads to relaxation and the formation of equilibrium reduced centers (curve 2). The same figure presents experimental points for native coupled phosphorylating mitochondria in the state of respiratory control, when the electron transfer rate is determined by ATP phosphorylation. Experimental points lie on the I(T) curve for the nonequilibrium center. After the uncoupling of the same preparation by the freezing and thawing method the phosphorylation reaction ceases to be the limiting state of electron transfer, and experimental points lie on the I(T) curve for the equilibrium N-2 center. The same effect is observed either after uncoupler addition [5.31, 33], or without an uncoupler if ADP and P_i concentrations are increased, i.e., increasing the rate of phosphorylation reaction [5.35] so that the overall process is now determined by the acts of electron transfer.

Nonequilibrium states of N-2 centers have also been recorded in our laboratory under conditions of reversed electron flow provided for by the ATP hydrolysis.

Experiments of HIND and JAGENDORF [5.36] (Sect.2.4) show that in the absence of ADP and P_i an undamaged system of membrane phosphorylation in chloroplasts can accumulate energy large enough to synthesize scores of ATP molecule per one ETC after the electron transfer is stopped and the phosphorylation substrates added. It can be assumed that in the coupled ETC of thylakoids, in the absence of phosphorylation substrates, the lifetime of carrier-transformers in nonequilibrium states increases so that between electron transfer acts the nonequilibrium state can spread to the surrounding lipoprotein membrane, whose relaxation time is even greater. In this way a thylakoid membrane can accumulate large quantities of energy and store them for a long time (especially at low temperatures). After the addition of phosphorylation substrates, ATP-synthetase complex (ATP-synthetase and carrier-transformer) relaxes during ATP synthesis, but the surrounding membrane again "pulls the trigger" and returns the complex into a "working" state. In this case the membrane plays the role of a high-capacity condenser in a scheme presented in Fig.5.8 [5.31].

In this scheme a low-capacity condenser C_1 represents a carrier-transformer, and a high-capacity condenser C_2 a membrane. The charging of the condenser is equivalent to the appearance of a strained conformationally nonequilibrium state of the carrier-transformer (for C_1) or of the membrane (for C_2). Key K_2 is switched on after C_1 is charged by phosphorylation substrates. Potential drop on the load R represents ATP synthesis, and generator E, ETC. Electron transfer is switched on by the K_1 key (in the case of photophosphorylation it is equivalent to illumination). The following conditions are fulfilled: $C_2 \gg C_1$; $\mathcal{T}_1 = R_1 C_1 < \mathcal{T}_2 = RC_1 < \mathcal{T}_3 = R_2 C_2$. In the presence of phosphorylation substrates, K_2 is switched on, and the discharge of C_1 ensures ATP synthesis owing to the energy liberated during electron transfer. If E is functioning (K_1 is switched on, and electron transfer is taking place), but phosphorylation substrates are absent (K_2 is switched off), only after C_1 has been charged does the charging of C_2 start (membrane transition into a nonequilibrium state). If phosphorylation substrates are now added (K_2 is switched on), and K_1 is switched off (e.g., the illumination of chloroplasts is switched off), C_1 will be discharged through R (ATP synthesis) and recharged from C_2 owing to the energy liberated during membrane relaxation.

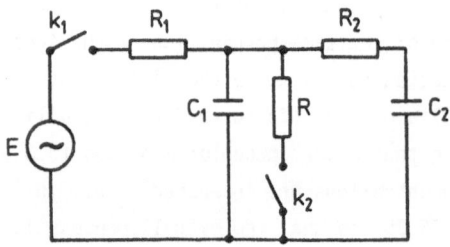

Fig.5.8. Equivalent scheme of ATP-synthetase in a membrane (after [5.31])

We have seen in Sect.3.4.3 that mitochondrial phosphorylation can take place in the absence of electron transport owing to a sufficiently fast change in the ionization state of groups with pH values between 8.1 and 8.3. In the case of mitochondria these are probably the groups in the carrier-transformer and/or ATP-synthetase. In the realm of relaxation concept it means that the ionization of these groups leads to the appearance of a conformationally nonequilibrium state, whose characteristics and relaxation path are similar to those of the nonequilibrium state appearing after the electron reduction of the carrier-transformer active center. With acid-base transition in experiments of JAGENDORF and URIBE on chloroplasts [5.37], the synthesis of scores of ATP molecules per ETC means that in this case the nonequilibrium state can be spread on to the thylakoidal membrane resulting in the accumulation of energy liberated during the titration of acid groups. It cannot be excluded that similar mechanism based on the titration of the ATP synthetase acid groups can explain the results of the beautiful experiments on photoinduced ATP synthesis in phospholipid vesicles with incorporated bacteriorodopsin and ATP-synthetase [5.38].

The thermodynamics of membrane phosphorylation was discussed in Sect.2.4 in the realm of ideas of classical physical chemistry. The experimental data and their theoretical analysis presented in this monograph show that these notions cannot serve as a basis for understanding the physical mechanisms involved. What is the relationship between the approach of the relaxation concept in bioenergetics (as well as of any "machine" concept which does not use the notions of dynamic equilibrium for the description of energy transduction acts) and the well known thermodynamic characteristics of ETC electron transport and ATP synthesis?

We have seen that under various conditions of membrane phosphorylation, with various ratios of phosphate potential to, e.g., electron flow through mitochondrial ETC, the stoichiometry of the "act per act" type between electron transfer and ATP formation is preserved. This means that the second term in (2.19) type does not affect the possibility of ATP synthesis as a result of redox reaction at a coupling site. However, under conditions of thermodynamic equilibrium between initial and end products of overall reaction, i.e., between NADH, O_2, ADP and P_i, on the one hand, and NAD^+ and ATP, on the other hand, the ATP synthesis is impossible. Thermodynamic equilibrium in its turn, presupposes the equality between the total free-energy values of the initial and the end products, i.e., depends on their concentrations. This apparent contradiction is easily resolved. In the coupled intact mitochondria under conditions of thermodynamic equilibrium both processes (electron transfer and phosphorylation) come to a stop. The respiratory control values for "good" preparations of tightly coupled mitochondria may exceed 40 [5.40]. This means that in coupled mitochondria the redox reaction at the coupled site cannot be realized without ADP phosphorylation: both processes proceed in one and the same elementary act. Under conditions of zero-free energy change this overall process simply comes to a stop: there proceed neither an "energy-donating" redox reaction nor an "energy-accepting" acid-base reaction of ADP and P_i condensation. The aforesaid does not forbid, of course, the formation of long-living intermediary nonequilibrium states of a carrier-transformer, an ATP-synthetase, and a membrane.

Let us now discuss the thermodynamic characteristics of a process which is realized in a coupling site and functions, to a considerable degree, as an individual molecular machine, in which electron transfer occurs between pairs of carriers fixed within ETC. This means that concentrations (activities) of donor and acceptor in a coupling site are equal, and that parameters determining the transduction process are not fully thermodynamic functions but their standard values.

Table 5.1 presents corresponding parameters for the NADH and succinate oxidation in mitochondria [5.41].

Table 5.1. Parameters for the NADH and succinate oxidation in mitochondria

Parameter	NADH oxidation	Succinate oxidation
ΔG^0 [kcal/mol]	-61.91	-36.2
ΔH^0 [kcal/mol]	-61.6	-36.2
ΔS^0 [kcal/mol K]	1.0 ± 4.4	-0.2 ± 4.0

It can be seen that entropy does not change in the course of these processes, and the reactions standard enthalpy values coincide with corresponding standard free-energy values. If we assume, in a first approximation, that an electron loses energy during its transfer along ETC only at coupling sites [5.42], then the transfer of one electron through a carrier-transformer on the path from NADH to O_2 is accompanied by the liberation of 10 kcal/mol, and on the path from succinate to O_2 9 kcal/mol. The standart enthalpy (as well as standard free-energy) value of an ATP formation under these conditions is 7-8 kcal/mol. In principle, thus with sufficiently high efficiency of the machine, the passage of one electron through a coupling site could ensure the synthesis of one ATP molecule. This "one ATP per one elect ron" ratio cannot be, of course realized simultaneously at all coupling sites. It would be desirable to carry out on mitochondria the experiments of the type which have been performed on chloroplasts with the light flashes of microsecond duration (Sect.3.4.2).

Theoretical considerations concerning mechanism of membrane phosphorylation similar to those presented in this chapter were formulated by CARTLING and EHRENBERG [5.43].

6. Conclusion

To conclude this monograph, which deals with the physical mechanisms of the most important bioenergetic processes, let me express some considerations of a more general nature. More than a hundred years ago in two different branches of science, certain new approaches were formulated almost simultaneously. These approaches got over the strong resistance of the inert mass of the established scientific concepts and became the basis for a new scientific Weltanschauung. I am referring to the works of Gibbs, who created statistical thermodynamics, and those of Darwin, who explained directional biological evolution by accidental hereditary changes fixed in accordance with the mechanism of natural selection. In both these approaches accidental elementary events ensure a strict regularity of the resultant macroscopic process. These approaches succeeded in time, though with difficulty, in overcoming the psychological barriers of mechanical determinism and have now become the only possible approaches for an over-whelming majority of scientists.

At the beginning of our century Van't Hoff and Arrhenius developed classical physical chemistry — a science of chemical equilibrium and chemical kinetics. This science is based on the idea of dynamic equilibrium: a priori probabilities of the rate constants of individual elementary acts of a chemical transformation in a state of chemical equilibrium are assumed to be the same as away from equilibrium. According to classical physical chemistry these probabilities are completely determined by the Boltzmann distribution of molecules. A similar approach determines the kinetics of chemical reactions in the realm of the Arrhenius and the activated state theories.

Although present-day chemists are often compelled to abandon the simplified schemes of classical physical chemistry in discussing the physical mechanisms of chemical transformations, this tendency practically did not touch the science of biochemistry. Theoretical analysis of experimental results concerning enzymatic catalysis, bioenergetic processes, etc., is almost always carried out on the basis of Arrhenius' and Van't Hoff's approaches. I think that it is becoming now an obstacle in the way of science. A partial return "backwards to mechanics", as any change in the scientific Weltanschauung, cannot, naturally, happen painlessly. Biochemistry has only recently obtained a theoretical basis with the help of classical physical chemistry, and it is extremely difficult to abandon the notion of its universality. I hope that this book may to some extent help in overcoming this psychological barrier.

References

Chapter 1

1.1 A. Szent-Györgyi: *Chemistry of Muscular Contraction* (Academic, New York 1947)
1.2 P. Mitchell: Nature **191**, 144 (1961)
1.3 L.A. Blumenfeld: *Problems of Biological Physics*, Springer Ser. Synergetics, Vol.7 (Springer, Berlin, Heidelberg, New York 1981)
1.4 L.A. Blumenfeld: *Problems of Biological Physics* (Nauka, Moscow 1974) in Russian

Chapter 2

2.1 J.R. Bendall: *Muscles, Molecules and Movement* (Heinemann, London 1969)
2.2 A.V. Hill: *First and Last Experiments in Muscle Mechanics* (University Press, Cambridge 1970)
2.3 V.I. Deshcherevsky: *Mathematical Models of Muscle Contraction* (Nauka, Moscow 1977) in Russian
2.4 A.F. Huxley: Prog. Biophys. and Bioph. Chem. **7**, 255 (1957)
2.5 H.E. Huxley: Cold Spring Harb. Symp. Quant. Biol. **37**, 361 (1973)
2.6 V.I. Deshcherevsky: Biorheology **7**, 147 (1971)
2.7 V.A. Engelhardt, M.N. Lubimova: Nature **144**, 668 (1939)
2.8 J.C. Seidel, J. Gergely: Biochem. Biophys. Res. Commun. **44**, 826 (1971)
2.9 J.C. Seidel, J. Gergely: Arch. Biochem. Biophys. **11**, 853 (1973)
2.10 M.M. Werber, A.G. Szent-Gyorgyu, G.D. Fasman: Biochem. **17**, 2872 (1972)
2.11 L.G. Ignatjeva, T.M. Seregina, L.A. Blumenfeld, E.K. Ruuge, R.I. Artjukh, G.B. Postnikova: Biofizika **17**, 533 (1972)
2.12 L.A. Blumenfeld, A.G. Ignatjeva: Eur. J. Biochem. **47**, 75 (1974)
2.13 E.K. Ruuge, N.V. Medvedeva, M.I. Vilenkina, L.A. Blumenfeld: Biofizika **21**, 409 (1976)
2.14 T. Yamata, H. Shimizu, H. Suga: Biochim. Biophys. Acta **305**, 642 (1973)
2.15 R.W. Lymn, E.W. Taylor: Biochem. **9**, 2975 (1970)
2.16 D.R. Trentham, R.C. Bardsley, J.F. Eccleston, A.G. Weeds: Biochem. J. **126**, 635 (1972)
2.17 C.R. Bagshaw, D.R. Trentham: Biochem. J. **133**, 323 (1973)
2.18 C.R. Bagshaw, D.R. Trentham: Biochem. J. **141**, 331 (1974)
2.19 C.R. Bagshaw, J.F. Eccleston, F. Eckstein, R.S. Goody, H. Gutfreund, D.R. Trentham: Biochem. J. **141**, 351 (1974)
2.20 C.R. Bagshaw, D.R. Trentham: J. Supramol. Struct. **3**, 315 (1975)
2.21 R.A. Alberty: J. Biol. Chem. **244**, 3290 (1969)
2.22 R.W. Lymn, E.W. Taylor: Biochem. **10**, 4617 (1971)
2.23 A.E. Bukatina, V.I. Deshcherevsky: Biofizika **17**, 738 (1972)
2.24 R.W. Limn: J. Theor. Biol. **43**, 313 (1974)
2.25 E. Esenberg, C. Moos: J. Biol. Chem. **245**, 2451 (1970)
2.26 H.G. Mannherz, H. Schenck, R.S. Goody: Eur. J. Biochem. **48**, 287 (1974)
2.27 N.P. Sidorenko, V.I. Deshcherevsky: Biofizika **15**, 785 (1970)
2.28 M.V. Volkenstein: Biochim. Biophys. Acta **180**, 562 (1969)
2.29 J.Ch. Skou: Biochim. Biophys. Acta **23**, 394 (1957)
2.30 J.Ch. Skou: Quart. Rev. Biophys. **7**, 401 (1974)

2.31 R. Whittam, A.R. Chipperfield: Biochim. Biophys. Acta **415**, 149 (1975)
2.32 Y. Kuriki, J. Halsey, R. Biltonen, E. Racker: Biochem. **15**, 4956 (1976)
2.33 A.A. Boldyrev, V.A. Tverdislov: *Molecular Organization and Functioning Mechanism of Na-Pump* (VINITI, Moscow 1978) in Russian
2.34 J.Ch. Skou: Biochim. Biophys. Acta **42**, 6 (1960)
2.35 S. Fahn, G. Koval, R. Albers: J. Biol. Chem. **241**, 1882 (1966)
2.36 R.L. Post, S. Kume, T. Tobin, B. Orcutt, A.K. Seu: J. Gen. Physiol. **54**, 306S (1969)
2.37 J.D. Robinson: FEBS Lett. **47**, 352 (1974)
2.38 A. Chipperfield, R. Whittam: Proc. Roy. Soc. A**187**, 269 (1974)
2.39 A. Schwartz, C.E. Lindenmayer, J.C. Allen: Farmacol. Rev. **27**, 3 (1975)
2.40 V.A. Tverdislov, L.V. Jakovenko: Biochimija **41**, 2088 (1976)
2.41 V.A. Tverdislov, L.V. Jakovenko, M.N. Resajeva: Mol. Biol. **13**, 377 (1979)
2.42 I.M. Glinn, J.F. Hoffmann, V.L. Lew: Phil. Trans. Roy. Soc. L.B. **262**, 91 (1971)
2.43 R. Post, K. Taniguchi, G. Toda: Abstracts of Intern. Conf. on Propert. Function Na, K-ATPase, New York (1973)
2.44 R. Post, G. Toda, S. Kume, K. Taniguchi: Pontif. Acad. Sci. Ser. Varia **40**, 341 (1976)
2.45 K. Taniguchi, R.L. Post: J. Biol. Chem. **250**, 3010 (1975)
2.46 R.L. Post: In *Biochemistry of Membrane Transport*, FEBS Symp. No.42, ed. by G. Semenza, E. Carafoli, Berlin (1977) p.352
2.47 E. Racker: *Mechanisms in Bioenergetics* (Academic, New York 1965)
2.48 D.E. Green, S. Fleisher: In *Horizons in Biochemistry*, ed. by M. Kasha, B. Pullman (Academic, New York 1962) p.381
2.49 B. Chance: FEBS Lett. **23**, 3 (1972)
2.50 J.S. Rieske, D.H. McLennan, R. Coleman: Biochem. Biophys. Res. Commun. **15**, 338 (1964)
2.51 B. Chance, C. Lee, L. Mela: Abstr. Symp. on Cytochromes, Osaka (1967) p.181
2.52 B.G. Malmström: Quart. Rev. Biophys. **6**, 389 (1973)
2.53 B.G. Malmström: Biochim. Biophys. Acta **549**, 281 (1979)
2.54 N.V. Vojevodskaja, D.Sh. Burbaev, A.F. Vanin, L.A. Blumenfeld: Mol. Biol. **15**, 243 (1981)
2.55 L.A. Blumenfeld: *Problems of Biological Physics*, Springer Ser. in Synergetics, Vol.7 (Springer, Berlin, Heidelberg, New York 1981)
2.56 M. Erecinska, D.F. Wilson, J.K. Blasiee: Biochim. Biophys. Acta **545**, 352 (1979)
2.57 J.C. Salerno, H. Blum, T. Ohnishi: Biochim. Biophys. Acta **547**, 270 (1979)
2.58 F.J. Ruzicka, H. Beinert, K.L. Schepter, W.R. Dunkann, R.H. Sands: Proc. Nat. Acad. Sci. USA **72**, 2886 (1975)
2.59 E.C. Slater: Quart. Rev. Biophys. **4**, 35 (1971)
2.60 S. Muraoka, E.C. Slater: Biochim. Biophys. Acta **180**, 221 (1969)
2.61 D. DeVault: J. Theor. Biol. **62**, 1115 (1976)
2.62 B. Chance, D.F. Wilson, P.L. Dutton, M. Erecinska: Proc. Nat. Acad. Sci. USA **66**, 1175 (1970)
2.63 N.A. Schroedl, C.R. Hartrell: Biochem. **16**, 1327 (1977)
2.64 N.A. Schroedl, C.R. Hartrell: Biochem. **16**, 4961 (1977)
2.65 N.A. Schroedl, C.R. Hartrell: Biochem. **16**, 4966 (1977)
2.66 T. Ohnishi: Eur. J. Biochem. **64**, 91 (1976)
2.67 J. Hatefi, W.G. Hanstein: Bioenergetics **3**, 129 (1972)
2.68 D.F. Wilson, P.L. Dutton: In *Electron and Coupled Energy Transfer in Biological Systems*, ed. by T.E. King, M. Klingenberg (Dekker, New York 1971) p.221
2.69 B. Chance, G.R. Williams: Adv. Enzymol. **17**, 65 (1956)
2.70 B. Chance, D. DeVault, L. Legalais, L. Mela, T. Jonetani: In *Fast Reactions and Primary Processes in Chemical Kinetics*, ed. by S. Claesson, 5th Nobel Symp. (Almqvist and Wiksell, Stockholm, 1967) p.437
2.71 B. Chance, C.P. Lee, T. Ohnishi, J. Higgins: In *Electron Transport and Energy Conservation*, ed. by J.M. Tager, S. Papa, E. Qungliariello, E.C. Slater (Adriatica Editrice, Bari 1970) p.29
2.72 L.P. Vernon: Bacteriol. Rev. **32**, 243 (1968)
2.73 M. Baltscheffsky: In *The Photosynthetic Bacteria*, ed. by R.K. Clayton, W.R. Sistrom (Plenum, New York 1978) p.595

2.74 A.L. Lehninger: *Biochemistry. The Molecular Basis of Cell Structure and Function* (Worth Publ., New York 1972)
2.75 A.Yu. Borisov, V.I. Godik: Biochim. Biopyhs. Acta **301**, 227 (1973)
2.76 M.G. Goldfeld, L.A. Blumenfeld: Bullet. Magn. Resonance **1**, 66 (1979)
2.77 H.T. Witt: In *Fast Reactions and Primary Processes in Chemical Kinetics*, ed. by S. Claesson, 5th Nobel Symp. (Almqvist and Wiksell, Stockholm 1967) p.261
2.78 H.T. Witt: Biochim. Biophys. Acta **505**, 355 (1979)
2.79 Y.K. Shen, G.M. Shen: Scientia Sin. **11**, 1097 (1962)
2.80 G. Hind, A.T. Jagendorf: Proc. Nat. Acad. Sci. USA **49**, 715 (1963)
2.81 E. Tyszkiewicz, E. Roux: Biochem. Biophys. Res. Commun. **65**, 1400 (1975)
2.82 A.T. Jagendorf: In *Bioenergetics of Photosynthesis*, ed. by Govindjee (Academic, New York 1975) p.413
2.83 N.Nelson, H. Nelson, E. Racker: J. Biol. Chem. **247**, 7657 (1972)
2.84 N. Lopez-Moratalla, E. Santiago, A.J. Iriarte, M.J. Lopez-Zabalza: Revista Espanola de Fisiologia **34**, 473 (1978)
2.85 E. Santiago, N. Lopez-Moratalla, J. Huaman, M.J. Lopez-Zabalza, A.J. Iriarte: Revista Espanola de Fisiologia **34**, 477 (1978)
2.86 E. Santiago, N. Lopez-Moratalla: Revista Espanola de Fisiologia **34**, 481 (1978)
2.87 E. Santiago, A.J. Iriarte, M.J. Lopez-Zabalza, N. Lopez-Moratalla: Arch. Biochem. Biophys. **196**, 1 (1979)
2.88 O.S. Nedelina, E.S. Vishnevsky, O.N. Brzhevskaja, E.M. Shekshejev, L.P. Kajushin: Dokl. Akad. Nauk SSSR **252**, 1497 (1980)
2.89 L.P. Kajushin, O.N. Brzhevskaja, O.S. Nedelina, E.M. Shekshejev: Biofizika **24**, 248 (1979)
2.90 M.G. Goldfeld, L.G. Dmitrovsky, L.A. Blumenfeld: Mol. Biol. **16**, 183 (1982)
2.91 W.G. Naustein: Biochem. Biophys. Acta **456**, 129 (1976)
2.92 The Mechanism of Energy Transduction in Biological Systems: Ann. NY Acad. Sci. **227**, 1 (1974)
2.93 C.W.F.McClare: Ann. NY Acad. Sci. **227**, 74 (1974)
2.94 Ann. NY Acad. Sci. **227**, 108, 179 (1974)

Chapter 3

3.1 E.C. Slater: Quart. Rev. Biophys. **4**, 35 (1971)
3.2 L.A. Blumenfeld: *Problems of Biological Physics*, Springer Ser. Synergetics, Vol.7 (Springer, Berlin, Heidelberg, New York 1981)
3.3 A.T. Jagendorf: In *Bioenergetics of Photosynthesis*, ed. by Govindjee (Academic, New York 1975) p.413
3.4 L. Ernster: Ann. Rev. Biochem. **46**, 981 (1977)
3.5 E.C. Slater: Ann. Rev. Biochem. **46**, 1015 (1977)
3.6 L.A. Blumenfeld: Quart. Rev. Biophys. **11**, 251 (1978)
3.7 N.M. Chernavskaja, D.S. Chernavsky: *Tunnel Transport of Electron in Photosynthesis* (University Press, Moscow 1977) in Russian
3.8 E.C. Slater: Nature **172**, 975 (1953)
3.9 M. Klingenberg: In *Biological Oxidations*, ed. by T.P. Singer (Interscience, New York 1968)p.3
3.10 G.I. Lichtenstein, A.E. Shilov: Izv. Akad. Nauk SSSR, Ser. Biol. 374 (1977)
3.11 B. Chance: FEBS Lett. **23**, 3 (1972)
3.12 L.A. Blumenfeld, M.I. Temkin: Biofizika **7**, 731 (1962)
3.13 A. Leninger: *The Mitochondrion* (Benjamin, New York 1964)
3.14 V.P. Skulachev: Nature **198**, 444 (1963)
3.15 V.P. Skulachev: Usp. Biol. Chim. **6**, 180 (1964)
3.16 V.P. Skulachev: *Energy Accumulation Within the Cell* (Nauka, Moscow 1969) in Russian
3.17 P. Mitchell: Nature **191**, 144 (1961)
3.18 P. Mitchell: Biol. Rev. **41**, 445 (1966)
3.19 P. Mitchell: FEBS Lett. **78**, 1 (1977)
3.20 P. Mitchell: Eur. J. Biochem. **95**, 1 (1979)
3.21 V.P. Skulachev: Usp. Sovr. Biol. **77**, 125 (1974)

3.22 A.T. Jagendorf: In *Photosynthesis. 1. Photosynthetic Electron Transport and Photophosphorylation*, ed. by A. Trebst, M. Avron (Springer, Berlin, Heidelberg, New York 1977) p.307

3.23 H.T. Witt: Biochim. Biophys. Acta **505**, 355 (1979)

3.24 V.P. Skulachev: *Energy Transformation within Biomembranes* (Nauka, Moscow 1972) in Russian

3.25 V.P. Skulachev: In *Energy Transducing Mechanisms*, ed. by E. Racker (Butterworth, London 1975) pp.31-73

3.26 D.B. Kell: Biochim. Biophys. Acta **549**, 55 (1979)

3.27 J.O.M. Bockris, A.K.N. Reddy: *Modern Electrochemistry* (Plenum, New York 1970)

3.28 B. Rumberg: In *Photosynthesis. 1. Photosynthetic Electron Transport and Photophosphorylation*, ed. by A. Trebst, M. Avron (Springer, Berlin, Heidelberg, New York 1977) p.405

3.29 R.J.P. Williams: J. Theor. Biol. **3**, 209 (1962)

3.30 R.J.P. Williams: Ann. NY Acad. Sci. **227**, 98 (1974)

3.31 R.J.P. Williams: Bioch. Soc. Trans. **1**, 1 (1973)

3.32 R.J.P. Williams: Proc. Roy. Soc. B**200**, 353 (1978)

3.33 R.J.P. Williams: Biochim. Biophys. Acta **505**, 1 (1978)

3.34 P. Mitchell: J. Bioenergetics **3**, 5 (1972)

3.35 E.S. Bauer: *Theoretical Biology* (VIEM Press, Moscow 1935) in Russian

3.36 H. Eyring, R. Lumry, J.D. Spikes: In *The Mechanism of Enzyme Action*, ed. by W. McElroy, B. Glass (J. Hopkins University Press, Baltimore 1954) p.123

3.37 Yu.I. Churgin, D.S. Chernavsky, S.E. Shnoll: Mol. Biol. **1**, 419 (1967)

3.38 S.E. Shnoll: In *Oscillatory Phenomena in Biological and Chemical Systems*, ed. by G.M. Frank (Nauka, Moscow 1967) p.22, in Russian

3.39 L. Packer: In Book of Abstracts, VI Intern. Congr. on Photobiology, Bochum (1972) N030

3.40 D.E. Green, S. Ji: Proc. Nat. Acad. Sci. USA **69**, 726 (1972)

3.41 D.E. Green: Biochim. Biophys. Acta **346**, 27 (1974)

3.42 D.E. Green: Ann. NY Acad. Sci. **227**, 108 (1974)

3.43 A. Bennun: Ann. NY Acad. Sci. **227**, 116 (1974)

3.44 A. Bennun: In *Proc. 3rd Intern. Congr. on Photosynthesis*, ed. by M. Avron (Elsevier, Amsterdam 1974) p.1107

3.45 A. Bennum: Biosystems **7**, 230 (1975)

3.46 P.D. Boyer, R.L. Cross, W. Momsen: Proc. Nat. Acad. Sci. USA **70**, 2837 (1973)

3.47 P.D. Boyer: Trends Biochem. Sci. **2**, 38 (1977)

3.48 S. Glasston, K.J. Laidler, H. Eyring: *Theory of Rate Processes. The Kinetics of Chemical Reactions, Viscosity, Diffusion, and Electrochemical Phenomena* (McGraw-Hill, New York 1941)

3.49 N.N. Semenov: *Chain Reactions* (Goskhimizdat, Leningrad 1934) in Russian

3.50 V.I. Vedeneev, O.M. Sarkisov, M.A. Totelboim, E.A. Shilov: Izv. Acad. Nauk SSSR, Chim. 1044 (1974)

3.51 A.N. Schilov: *On the Coupled Oxidation Reactions* (University Press, Moscow 1905) in Russian

3.52 L.A. Blumenfeld: Biofizika **21**, 946 (1976)

3.53 G. Weber: Ann. NY Acad. Sci. **227**, 486 (1974)

3.54 H.N. Christensen: J. Ther. Biol. **57**, 419 (1976)

3.55 E.A. Guggenheim: *Modern Thermodynamics by the Methods of Willard Gibbs* (Methuen, London 1933)

3.56 B. Lowenhaupt: Ann. NY Acad. Sci. **227**, 214 (1974)

3.57 H. Tedeschi: *Mitochondria: Structure, Biogenesis and Transducing Functions* (Springer, Wien, New York 1976)

3.58 P. Mitchell, J. Moyle: Eur. J. Biochem. **7**, 471 (1969)

3.59 D.G. Nichols: Eur. J. Biochem. **50**, 305 (1974)

3.60 N. Kamo, M. Muratsugu, R. Hongoh, Y. Kobatane: J. Mol. Biol. **49**, 105 (1979)

3.61 E.A. Liberman, V.P. Topali, L.M. Tsofina, A.A. Jasaitis, V.P. Skulachev: Nature **222**, 1076 (1969)

3.62 E.A. Liberman, V.P. Skulachev: Biochim. Biophys. Acta **216**, 30 (1970)

3.63 V.P. Skulachev: FEBS Lett. **11**, 301 (1970)

3.64 V.P. Skulachev: Current Topics Bioenergetics **4**, 127 (1971)

3.65 V.P. Skulachev: FEBS Lett. **87**, 171 (1978)

3.66 H. Tedeschi: FEBS Lett. **59**, 1 (1975)
3.67 J.Th.G. Overbeek: Progr. Biophys. Biophysical Chemistry **6**, 58 (1956)
3.68 D. Ort, R.A. Dilley: Biochim. Biophys. Acta **449**, 95 (1976)
3.69 J. Lavorel: In *Bioenergetics of Photosynthesis*, ed. by Govindjee (Academic, New York 1975) p.223
3.70 J. Barber, G.P.B. Kraan: Biochim. Biophys. Acta **197**, 49 (1970)
3.71 J. Barber: FEBS Lett. **20**, 251 (1972)
3.72 C.A. Wraight, A.R. Crofts: Eur. J. Biochem. **19**, 386 (1971)
3.73 W. Junge: Ann. Rev. Plant. Physiol. **28**, 503 (1977)
3.74 J.B. Jackson, A.R. Crofts: FEBS Lett. **4**, 185 (1969)
3.75 W. Junge, B. Rumberg, H. Schröder: Eur. J. Biochem. **14**, 575 (1970)
3.76 B. Rumberg, U. Siggel: Naturwissensch. **56**, 130 (1969)
3.77 W. Junge, H.T. Witt: Z. Naturforsch. **23**B, 244 (1968)
3.78 L.N.M. Duysens: Science **120**, 353 (1954)
3.79 H.T. Witt: Naturwissensch. **42**, 72 (1955)
3.80 S. Schmidt, R. Reich, H.T. Witt: Naturwissensch. **58**, 414 (1970)
3.81 M. Baltsheffsky: In *Dynamics of Energy-Transducing Membranes*, ed. by L. Ernster, R. Estabrook, E.C. Slater (Elsevier, Amsterdam 1974) p.365
3.82 M. Baltsheffsky, D.O. Hall: FEBS Lett. **39**, 345 (1974)
3.83 L.A. Blumenfeld, F.P. Chernyakovskii, V.A. Gribanov, I.M. Kanevskii: J. Macromolecular Sci.-Chemistry A**6**, 1201 (1972)
3.84 R.P. Feinmann: Phys. Rev. **56**, 340 (1939)
3.85 L.A. Drachev, A.A. Jasaitis, A.D. Kaulen, A.A. Kondrashin, E.A. Liberman, I.B. Nemecek, S.A. Ostroumov, A.Yu. Semenov, V.P. Skulachev: Nature **249**, 321 (1974)
3.86 A.Yu. Semenov, S.K. Chimarovskij, U.A. Smirnova, L.A. Drachev, A.A. Kononenko, H.Ya. Uspenskaja, A.B. Rubin, V.P. Skulachev: Mol. Biol. **15**, 622 (1981)
3.87 J.T. Tupper, R. Rikmenspool: Rev. Sci. Instr. **40**, 851 (1969)
3.88 E.A. Liberman, V.P. Skulachev: Biochim. Biophys. Acta **216**, 30 (1970)
3.89 J.T. Tupper, H. Tedeshi: Science **166**, 1539 (1969)
3.90 E.J. Harris, B.C. Pressman: Biochim. Biophys. Acta **172**, 66 (1969)
3.91 E.J. Harris, D.E. Bassett: FEBS Lett. **19**, 214 (1972)
3.92 A.A. Bulychev, V.K. Andrianov, G.A. Kurella, F.F. Litvin: Nature **236**, 175 (1972)
3.93 W.J. Vredenberg: In *Proc. 3rd Intern. Congr. on Photosynthesis*, ed. by M. Avron (Elsevier, Amsterdam 1974) p.929
3.94a C. Deutch, M. Erecinska, R. Werrlein, J.A. Silver: Proc. Nat. Acad. Sci. USA **76**, 2175 (1979)
3.94b D.F. Wilson, N.S. Forman: Biochemistry **21**, 1438 (1982)
3.95 R. Kraayenhof: Biochim. Biophys. Acta **180**, 213 (1969)
3.96 Ch. Giersch, U. Heber, Y. Kobayashi, V. Inone, K. Shibata, H.W. Heldt: Biochim. Biophys. Acta **590**, 59 (1980)
3.97 K.A. Santarius, U. Heber: Biochim. Biophys. Acta **102**, 39 (1965)
3.98 R. Lilley, C.J. Chou, A. Mosback, H.W. Heldt: Biochim. Biophys. Acta **460**, 259 (1977)
3.99 U. Heber, K.A. Santarius: Z. Naturforschg. **25**b, 718 (1970)
3.100 J.L. Bomsel, A. Pradet: Biochim. Biophys. Acta **162**, 230 (1968)
3.101 Y. Kobayashi, Y. Inone, F. Furuja, K. Shibata, U. Heber: Planta **147**, 69 (1979)
3.102 J.W. Stucki: In *Energy Conservation in Biological Membranes*, ed. by G. Schäfer, M. Klingenberg (Springer, Berlin, Heidelberg, New York 1978) p.264
3.103 S.P. Robinson, J.T. Wiskich: Biochim. Biophys. Acta **440**, 131 (1976)
3.104 U. Enser, U. Heber: Biochim. Biophys. Acta **592**, 577 (1980)
3.105 A.N. Tikhonov, G. Khomutov, E.K. Runge, L.A. Blumenfeld: Biochim. Biophys. Acta **637**, 321 (1981)
3.106 M. Avron: In Book of Abstracts, VI Internat. Congr. on Photobiology, Bochum (1972) No.035
3.107 J.J. Schurmans, R.P. Casey, R. Kraayenhof: FEBS Lett. **94**, 405 (1978)
3.108 H.T. Witt: In *Bioenergetics of Photosynthesis*, ed. by Govindjee (Academic, New York 1975) p.493

3.109 H.H. Grünhagen, H.T. Witt: Z. Naturforschg. **25**B, 373 (1970)
3.110 H.T. Witt: Quart. Rev. Biophys. **4**, 365 (1971)
3.111 L.A. Blumenfeld, M.G. Goldfeld, L.G. Dmitrovsky: Stud. Biophys. **65**, 69 (1977)
3.112 M.G. Goldfeld, L.G. Dmitrovsky, L.A. Blumenfeld: Biofizika **22**, 357 (1977)
3.113 M.G. Goldfeld, L.G. Dmitrovsky, L.A. Blumenfeld: Mol. Biol. **12**, 857 (1978)
3.114 P. Gräber, H.T. Witt: Biochim. Biophys. Acta **423**, 141 (1976)
3.115 S. Del Va-le-Tascon, R. van Groudelle, L.N.M. Duysens: Biochim. Biophys. Acta **504**, 26 (1978)
3.116 R.M. Petty, J.B. Jackson: Biochim. Biophys. Acta **547**, 474 (1979)
3.117 D.A. Harris, M. Baltsheffsky: Biochem. Biophys. Res. Commun. **86**, 1248 (1979)
3.118 S.J. Ferguson, P. John, W.J. Lloyd, G.K. Radda, F.R. Whatley: FEBS Lett. **62**, 272 (1976)
3.119 H. Komai, D.R. Hunter, Y. Takahashi: Biochem. Biophys. Res. Commun. **53**, 82 (1973)
3.120 D.R. Hunter, H. Komai, R.A. Haworth: Biochem. Biopyhs. Res. Commun. **56**, 647 (1974)
3.121 H. Komai, D.R. Hunter, J.H. Southard, R.A. Haworth, D.E. Green: Biochem. Biophys. Res. Commun. **69**, 695 (1976)
3.122 A.T. Jagendorf, E. Uribe: Proc. Nat. Acad. Sci. USA **55**, 170 (1966)
3.123 T. Yamamoto, Y. Tonomura: J. Biochem. (Tokio) **77**, 137 (1975)
3.124 J.M. Gould, L.K. Patterson, E. Ling, G.D. Winget: Nature **280**, 607 (1979)
3.125 R.P. Magnusson, R.E. McCarty: J. Biol. Chem. **251**, 6874 (1976)
3.126 M.G. Goldfeld, L.G. Dmitrovsky, L.A. Blumenfeld: Biofizika **23**, 549 (1978)
3.127 R.A. Reid, J. Moyle, P. Mitchell: Nature **212**, 257 (1966)
3.128 I.V. Malenkova, S.P. Kuprin, R.M. Davidov, L.A. Blumenfeld: Dokl. Acad. Nauk SSSR **252**, 743 (1980)
3.129 I.V. Malenkova, S.P. Kuprin, R.M. Davidov, L.A. Blumenfeld: Biochim. Biophys. Acta **682**, 179 (1982)
3.130 H.T. Witt, E. Schlodder, P. Graber: FEBS Lett. **69**, 272 (1976)
3.131 T. Hamamoto, K. Ohno, Y. Kagawa: J. Biochem. (Tokyo) **91**, 1759 (1982)
3.132 M. Rogner, K. Ohno, T. Hamamoto, N. Sone, Y. Kagawa: Biochem. Biophys. Res. Commun. **91**, 362 (1979)
3.133 Ch. Vincler, R. Korenstein: Proc. Nat. Acad. Sci. USA **79**, 3183 (1981)
3.134 J.H. Wang, M. Jang: Biochem. Biopyhs. Res. Commun. **73**, 673 (1976)
3.135 S.V. Konev, A.N. Rasumovitch, N.V. Gamezo, A.N. Rudenok: Biofizika **24**, 349 (1979)

Chapter 4

4.1   L.A. Blumenfeld: Biofizika **16**, 724 (1971)
4.2   L.A. Blumenfeld: *Problems of Biological Physics*, Springer Ser. Synergetics, Vol.7 (Springer, Berlin, Heidelberg, New York 1981)
4.3   E.C. Slater: Quart. Rev. Biophys. **4**, 35 (1971)
4.4   L.A. Blumenfeld: Biofizika **17**, 954 (1972)
4.5   L.A. Blumenfeld, V.K. Koltover: Mol. Biol. **6**, 161 (1972)
4.6   B.F. Gray: Nature **253**, 436 (1975)
4.7   L. Brillouin: *Science and Information Theory* (Academic, New York 1956)
4.8   R.P. Feinmann: Phys. Rev. **56**, 430 (1939)
4.9   B.F. Gray, I. Gonda: J. Theor. Biol. **69**, 167 (1977)
4.10 L.A. Blumenfeld, V.I. Goldansky, M.I. Podgoretzky, D.S. Chernavsky: Zh. Strukt. Chimii **8**, 854 (1967)
4.11 I.M. Lifshitz: J. Exp. Theor. Phys. **55**, 2408 (1986)
4.12 I.M. Lifshitz, A.Yu. Grosberg, A.R. Khokhlov: Rev. Mod. Phys. **50**, 683 (1978)
4.13 Q.H. Gibson: Biochem. J. **71**, 293 (1959)
4.14 B. Alpert, R. Banerjee, L. Lindqvist: Proc. Nat. Acad. Sci. USA **71**, 558 (1974)
4.15 L.A. Blumenfeld, R.M. Davydov, S.N. Magonov, R.O. Vilu: FEBS Lett. **49**, 246 (1974)
4.16 L.A. Blumenfeld, D.Sh. Burbaev, A.F. Vanin, R.O. Vilu, R.M. Davydov, S.N. Magonov: Zh. Strukt. Chimii **15**, 1030 (1974)
4.17 L.A. Blumenfeld, R.M. Davydov, N.S. Fel', S.N. Magonov, R.O. Vilu: FEBS Lett. **45**, 256 (1974)

4.18 S.N. Magonov, L.A. Blumenfeld, V.K. Vanag, R.M. Davydov: Biofizika **23**, 414 (1978)

4.19 R.M. Davydov, S.N. Magonov, A.M. Arutjunjan, Yu.A. Sharonov: Mol. Biol. **12**, 1341 (1978)

4.20 L.A. Blumenfeld, R.M. Davydov, R.O. Vilu, S.M. Magonov: In IFIAS Workshop on Physicochemical Aspects of Electron Transfer Processes in Enzyme Systems. IFIAS Stockholm (1977) p.16

4.21 F. Heinmets: Physiol. Chem. Phys. **10**, 1 (1978)

4.22 R. Swanson, B.L. Trus, N. Mandel, G. Mandel, O.B. Kallai, R.E. Dickerson: J. Biol. Chem. **252**, 759 (1977)

4.23 T. Takano, B.L. Trus, N. Mandel, G. Mandel, O.B. Kallai, R. Swanson, R.E. Dickerson: J. Biol. Chem. **256**, 776 (1977)

4.24 R.M. Davydov, A.B. Karyakin, S. Greschner: Biofizika **25**, 393 (1980)

4.25 M. Perutz, L.F. Ten Euck: Cold. Spring Harbor Symp. Quant. Biol. **36**,295 (1971)

4.26 E.J. Heidner, R.C. Ladner, M.F. Perutz: J. Mol. Biol. **104**, 707 (1976)

4.27 I.B. Bersucker, S.S. Stavrov, B.G. Vechter: Biofizika **24**, 413 (1979)

4.28 S.N. Magonov, R.M. Davydov, L.A. Blumenfeld, R.O. Vilu, A.M. Arutjunjan, Yu.A. Sharonov: Mol. Biol. **12**, 947 (1978)

4.29 L.A. Blumenfeld, S.N. Magonov, A.M. Arutjunjan, R.M. Davydov, Yu.A. Sharonov: In *III USSR Symp. on Biopolymer Spectroscopy, Reports*, ed. by V.Ya Maleev, Kharkov (1977) p.13

4.30 R.M. Davydov: Biofizika **25**, 203 (1980)

4.31 A.M. Arutjunjan, S.N. Magonov, Yu.A. Sharonov: Mol. Biol. **13**, 438 (1979)

4.32 S.N. Magonov, R.M. Davydov, L.A. Blumenfeld, R.O. Vilu, A.M. Arutjunjan, Yu.A. Sharonov: Mol. Biol. **12**, 1182 (1978)

4.33 S.N. Magonov, A.M. Arutjunjan, L.A. Blumenfeld, R.M. Davydov, Yu.A. Sharonov: Dokl. Akad. Nauk SSSR **232**, 695 (1978)

4.34 S.N. Magonov, R.M. Davydov, L.A. Blumenfeld, A.M. Arutjunjan, Yu.A. Sharonov: Mol. Biol. **12**, 1191 (1978)

4.35 R.M. Davydov, S. Greschner, G. Ruckpaul: Mol. Biol. **13**, 1397 (1979)

4.36 S. Greschner, R.M. Davydov, G.R. Jänig, K. Ruckpaul, L.A. Blumenfeld: Acta Biol. Med. Ger. **38**, 443 (1979)

4.37 R.M. Davydov: Mol. Biol. **14**, 272 (1980)

4.38 J.P. Ganda, J.F. Gibson, R. Cammack, D.O. Hall, R. Mullinger: Biochim. Biophys. Acta **434**, 154 (1976)

4.39 L.A. Blumenfeld, D.Sh. Burbaev, R.M. Davydov, L.N. Kubrina, A.F. Vanin, R.O. Vilu: Biochim. Biophys. Acta **379**, 512 (1975)

4.40 J.E. Wertz, J.R. Bolton: *Electron Spin Resonance, Elementary Theory and Practical Applications* (McGraw-Hill, New York 1972)

4.41 Ch.W. Carter, Jr., J. Kraut, S.T. Freer, R.A. Alden: J. Biol. Chem. **249**, 6339 (1974)

4.42 L.A. Blumenfeld, D.Sh. Burbaev, A.V. Lebanidze, A.F. Vanin: Stud. Biophys. **63**, 143 (1977)

4.43 A.V. Lebanidze, D.Sh. Burbaev, L.A. Blumenfeld: Zh. Strukt. Chim. **19**, 448 (1978)

4.44 D.Sh. Burbaev, A.V. Lebanidze: Biofizika **24**, 392 (1979)

4.45 D.Sh. Burbaev, A.V. Lebanidze: Biofizika **24**, 552 (1979)

4.46 E.J. Land, A.J. Swallow: Arch. Biochem. Biophys. **145**, 365 (1971)

4.47 I. Pecht, M. Faraggi: Proc. Nat. Acad. Sci. USA **69**, 902 (1972)

4.48 M. Faraggi, I. Pecht: Isr. J. Chem. **10**, 1021 (1972)

4.49 J. Wilting, R. Braams, H. Nauta, K.J.H. van Buuren: Biochim. Biophys. Acta **283**, 543 (1972)

4.50 N.N. Lichtin, A. Shafferman, G. Stein: Biochim. Biophys. Acta **314**, 117 (1973)

4.51 N.N. Lichtin, A. Shafferman, G. Stein: Biochim. Biophys. Acta **357**, 386 (1974)

4.52 E.J. Land, A.J. Swallow: Biochim. Biophys. Acta **368**, 89 (1974)

4.53 A. Shafferman, G. Stein: Biochim. Biophys. Acta **416**, 287 (1975)

4.54 N.S. Fel', P.I. Dolin, R.M. Davydov, V.K. Vanag, S.P. Kuprin, L.A. Blumenfeld: Elektrochimia **13**, 909 (1977)

4.55 L.A. Blumenfeld, R.M. Davydov, S.P. Kuprin, S.V. Stepanov: Biofizika **22**, 977 (1977)

4.56 S.P. Kuprin, R.M. Davydov, N.S. Fel', R.M. Nalbandjan, L.A. Blumenfeld:
      Dokl. Akad. Nauk SSSR **235**, 1193 (1977)
4.57 R.M. Davydov, S.P. Kuprin: Biofizika **26**, 150 (1981)
4.58 R.M. Davydov, S.P. Kuprin, N.S. Fel', N.F. Nesnayko, E.N. Mukhin, L.A.
      Blumenfeld: Dokl. Akad. Nauk SSSR **239**, 220 (1978)
4.59 L.S. Kaminsky, V.J. Miller, A.J. Davison: Biochem. **12**, 2215 (1973)
4.60 M.G. Simis, I.A. Taub: Biophys. J. **24**, 285 (1978)
4.61 R.M. Davydov, S.V. Stepanov: Biofizika **25**, 624 (1980)
4.62 T. Brittain, M.T. Wilson, C. Greenwood: Biochem. J. **141**, 455 (1974)
4.63 T. Brittain, C. Greenwood: Biochem. J. **149**, 713 (1975)
4.64 S. Dmitrov, R.M. Davydov, S.P. Kuprin: Biofizika **26**, 386 (1981)
4.65 C.A. Sawicki, Q.H. Gibson: J. Biol. Chem. **251**, 1533 (1976)
4.66 J. Friedman, R.B. Lyons, P.A. Fleury: Biophys. J. **25**, 128a (1979)
4.67 L.J. Noe, W.G. Eisert, P.M. Rentzepis: Biophys. J. **24**, 379 (1978)
4.68 R.M. Davydov, S.V. Stepanov, A.P. Ledenev, L.A. Blumenfeld: Dokl. Akad.
      Nauk SSR **242**, 965 (1978)
4.69 R.M. Davydov, S. Greshner, S.V. Stepanov, K. Ruckpaul: Mol. Biol. **14**, 685
      (1980)
4.70 S.V. Stepanov, R.M. Davydov, Biofizika **25**, 364 (1980)
4.71 S. Greschner, L.A. Blumenfeld, M.V. Genkin, R.M. Davydov, N.M. Roldugina:
      Stud. Biophys. **57**, 109 (1976)
4.72 L.A. Blumenfeld, S. Greschner, M.V. Genkin, R.M. Davydov, N.M. Rodlugina:
      Stud. Biophys. **57**, 110 (1976)
4.73 L.A. Blumenfeld, R.M. Davydov: Biochim. Biophys. Acta **549**, 225 (1979)
4.74 L.A. Blumenfeld: In *Molecular Interactions and Activity in Proteins*, Ciba
      Foundation Symp. 60 (New Series), Excerpta Med. London (1978) p.47
4.75 R.M. Davydov, S.P. Kuprin, N.S. Fel', G.B. Postnikova, L.A. Blumenfeld:
      Dokl. Akad. Nauk SSSR **235**, 950 (1977)
4.76 V.V. Kuprijanov, A.S. Pobotchin, V.N. Lusikov: Biochimia **41**, 1889 (1976)
4.77 Yu.A. Ermakov, V.I. Pasechnik, S.V. Tulsky: Biofizika **21**, 629 (1976)
4.78 V.I. Pasechnik: Biofizika **21**, 746 (1976)
4.79 Yu.A. Ermakov, S.P. Kuprin, V.I. Pasechnik: Biofizika **21**, 788 (1976)
4.80 L.A. Blumenfeld, Yu.A. Ermakov, V.I. Pasechnik: Biofizika **22**, 8 (1977)
4.81 L.A. Blumenfeld, Yu.A. Ermakov, V.I. Pasechnik: Biofizika **22**, 535 (1977)
4.82 E. Antonini, M. Brunori, A. Colosimo, C. Greenwood, M.T. Wilson: Proc.
      Nat. Acad. Sci. USA **74**, 3128 (1977)
4.83 S. Rosen, R. Bränden, T. Vänngard, B.G. Malmström: FEBS Lett. **74**, 25 (1977)
4.84 D.Sh. Burbaev, A.V. Lebanidze: Biofizika **21**, 942 (1976)

Chapter 5

5.1   L.A. Blumenfeld, V.K. Koltover: Mol. Biol. **6**, 161 (1972)
5.2   M.F. Morales, J. Boots: Proc. Nat. Acad. Sci. USA **78**, 3857 (1979)
5.3   H.M. Levy, F. Ramizer, R.K. Shukla: J. Teor. Biol. **81**, 327 (1979)
5.4   D.E. Koshland, Jr.: Proc. Nat. Acad. Sci. USA **44**, 98 (1958)
5.5   V.V. Lednev, A.N. Konev, L.K. Srebnitzkaja, S.B. Malinchik, A.S. Khromov:
      Biofizika **27**, N6 (1982)
5.6   W.F. Harrington: Proc. Nat. Acad. Sci. USA **68**, 685 (1971)
5.7   N.F. Kastrikin: J. Theor. Biol. **84**, 387 (1980)
5.8   E.W. Taylor: CRC Crit. Rev. Biochem. **6**, 103 (1979)
5.9   B.F. Gray, I. Gonda: J. Theor. Biol. **69**, 167 (1977)
5.10 B.F. Gray, I. Gonda: J. Theor. Biol. **69**, 187 (1977)
5.11 I. Gonda, B.F. Gray: In *Biomolecular Structure, Conformation, Function and
      Evolution*, Vol.2 (Pergamon, Oxford 1980) p.609
5.12 R.P. Feinmann: Phys. Rev. **56**, 340 (1939)
5.13 B.F. Gray, I. Gonda: J. Theor. Biol. **49**, 493 (1975)
5.14 B.F. Gray: Nature **253**, 436 (1975)
5.15 C.W.F. McClare: J. Theor. Biol. **35**, 569 (1972)
5.16 H. Shimizu: J. Phys. Soc. Japan **32**, 1323 (1972)
5.17 O. Jardetzky: Nature **211**, 969 (1966)

5.18 W.D. Stein, V. Eilam, W.R. Lieb: Ann. NY Acad. Sci. **227**, 328 (1974)
5.19 D.C. Levitt: Biochim. Biophys. Acta **604**, 321 (1980)
5.20 A.A. Boldyrev, V.A. Tverdislov: *Molecular Organization and Action Mechanism of Na-Pump* (VINITI, Moscow 1978) in Russian
5.21 V.A. Tverdislov, L.V. Yakovenko, M.N. Resaeva: Mol. Biol. **13**, 377 (1979)
5.22 V.A. Tverdislov, M.N. Resaeva, A.N. Tikhonov, V.I. Lobyshev: Mol. Biol. **14**, 1362 (1980)
5.23 V.A. Tverdislov, L.V. Yakovenko: Biofizika **25**, 815 (1980)
5.24 L.A. Blumenfeld: *Problems of Biological Physics*, Springer Ser. Synergetics, Vol.7 (Springer, Berlin, Heidelberg, New York 1981)
5.25 L.A. Blumenfeld: Biofizika **17**, 954 (1972)
5.26 L.A. Blumenfeld: J. Theor. Biol. **58**, 269 (1976)
5.27 N.P. Sidorenko, V.I. Deshcherevsky: Biofizika **15**, 785 (1970)
5.28 Yu.A. Chismadzhev, V.F. Pastushenko, L.A. Blumenfeld: Biofizika **21**, 208 (1976)
5.29 V.M. Fain: J. Chem. Phys. **65**, 1854 (1976)
5.30 I.M. Lifshitz: J. Exp. Theor. Phys. **55**, 2408 (1968)
5.31 L.A. Blumenfeld: Quart. Rev. Biophys. **11**, 251 (1978)
5.32 L.A. Blumenfeld, D.S. Chernavskii: J. Theor. Biol. **39**, 1 (1973)
5.33 L.A. Blumenfeld, D.Sh. Burbaev, A.V. Lebanidze: In *Magnetic Resonance in Biology and Medicine*. Theses of 1st Symp., ed. by N.M. Emanuel (Nauka, Moscow 1977) p.82, in Russian
5.34 D.Sh. Burbaev: Biofizika **24**, 1099 (1979)
5.35 D.Sh. Burbaev, I.P. Solozhenkin, L.A. Blumenfeld: In *Magnetic Resonance in Biology and Medicine*. Theses of the 2nd Symp., ed. by N.M. Emanuel (Nauka, Moscow 1981) p.128, in Russian
5.36 G. Hind, A.T. Jagendorf: Proc. Nat. Acad. Sci. USA **49**, 715 (1963)
5.37 A.T. Jagendorf, E. Uribe: Proc. Nat. Acad. Sci. USA **55**, 170 (1966)
5.38 G.D. Winget, N. Kanner, E. Racker: Biochim. Biophys. Acta **460**, 490 (1977)
5.39 W.F. Harrington: Proc. Nat. Acad. Sci. USA **76**, 5066 (1979)
5.40 B. Chance, G.R. Williams: Advan. Enzymol. **17**, 65 (1956)
5.41 M. Poe, H. Gutfreund, R.M. Estabrook: Arch. Biochem. Biophys. **122**, 204 (1967)
5.42 B. Chance: FEBS Lett. **23**, 3 (1972)
5.43 B. Cartling, A. Ehrenberg: Biophys. J. **23**, 451 (1978)

Subject Index

C. P. Slichter

Principles of Magnetic Resonance

1980. 115 figures, 9 tables. XII, 397 pages. (Springer Series in Solid-State Sciences, Volume 1). ISBN 3-540-08476-2

Contents: Elements of Resonance. – Basic Theory. – Magnetic Dipolar Broadening of Rigid Lattices. – Magnetic Interactions of Nuclei with Electrons. – Spin-Lattice Relaxation and Motional Narrowing of Resonance Lines. – Spin Temperature in Magnetism and in Magnetic Resonance. – Double Resonance. – Advanced Concepts in Pulsed Magnetic Resonance. – Electric Quadrupole Effects. – Electron Spin Resonance. – Summary. – Problems. – Appendices. – Selected Bibliography. – References. – Autor Index.– Subject Index.

From the reviews: "... the book remains one of the best expositions of the quantum theory of resonance in existence. It has my highest recommendation for use as a textbook." *Journal of the Optical Society of America*

"... an important introductory book... Slichter, professor of physics and his graduate students... are responsible for numerous original and significant research contributions that appear in the appear in the book. The clarity and style in which the book is written reveals Slichter's research expertise and talent as an excellent teacher and expositor... The referencing is so good that certain new priorities in research contributions to NMR appear that were not previously obvious..." *Physics Today*

"... A comment on Professor Slichter's pedagogical technique is in order. He picks a simple example elucidating the main physical features of a subject, and treats it in detail. Thus one has a feel application of the general formalism to more complex problems. On the whole, this is an excellent updating of what was already an established standard in the field. It is particularly timely since commercial equipment has recently become available enabling researchers to perform some of these newer experiments without tedious apparatur development." *Applied Spectroscopy*

Mössbauer Spectroscopy

Editor: U. Gonser
1975. 96 figures. XVIII, 241 pages. (Topics in Applied Physics, Volume 5) ISBN 3-540-07120-2

Contents: *U. Gonser:* From a Strange Effect to Mössbauer Spectroscopy. – *P. Gütlich:* Mössbauer Spectroscopy in Chemistry. – *R. W. Grant:* Mössbauer Spectroscopy in Magnetism: Characterization of Magnetically-Ordered Compounds. – *C. E. Johnson:* Mössbauer Spectroscopy in Biology. – *S. S. Hafner:* Mössbauer Spectroscopy in Lunar Geology and Mineralogy. – *F. E. Fujita:* Mössbauer Spectroscopy in Physical Metallurgy.

Mössbauer Spectroscopy II

The Exotic Side of the Method
Editor: U. Gonser
1981. 67 figures. XII, 196 pages. (Topics in Current Physics, Volume 25) ISBN 3-540-10519-0

Contents: *U. Gonser:* Introduction. – *R. L. Mössbauer, F. Parak, W. Hoppe:* A Solution of the Phase Problem in the Structure Determination of Biological Macromolecules. – *R. V. Pound:* The Gravitational Red-Shift. – *V. I. Goldanskii, R. N. Kuzmin, V. A. Namiot:* Trends in the Development of the Gamma Laser. – *R. L. Cohen:* Nuclear Resonance Experiments Using Synchrotron Sources. – *U. Gonser, H. Fischer:* Resonance γ-Ray Polarimetry. – *B. D. Sawicka, J. A. Sawicki:* Iron-Ion Implantation Studied by Conversion-Electron Mössbauer Spectroscopy. – *R. S. Preston, U. Gonser:* Selected "Exotic" Applications. – *S. S. Hanna:* The Discovery of the Magnetic Hyperfine Interaction in the Mössbauer Effect of ^{57}Fe.

Springer-Verlag
Berlin
Heidelberg
New York
Tokyo

Hydrodynamic Instabilities and the Transition to Turbulence

Editors: **H. L. Swinney, J. P. Gollub**

1981. 81 figures. XII, 292 pages
(Topics in Applied Physics, Volume 45)
ISBN 3-540-10390-2

Contents: *H. L. Swinney, J. P. Gollub:* Introduction. – *O. E. Lanford:* Strange Attractors and Turbulence. – *D. D. Joseph:* Hydrodynamic Stability and Bifurcation. – *J. A. Yorke, E. D. Yorke:* Chaotic Behavior and Fluid Dynamics. – *F. H. Busse:* Transition to Turbulence in Rayleigh-Bénard Convection. – *R. C. DiPrima, H. L. Swinney:* Instabilities and Transition in Flow Between Concentric Rotating Cylinders. – *S. A. Maslowe:* Shear Flow Instabilities and Transition. – *D. J. Tritton, P. A. Davies:* Instabilities in Geophysical Fluid Dynamics. – *J. M. Guckenheimer:* Instabilities and Chaos in Nonhydrodynamic Systems.

J. Schnakenberg

Thermodynamic Network Analysis of Biological Systems

Universitext

2nd corrected and updated edition. 1981. 14 figures. X, 149 pages
ISBN 3-540-10612-X

Contents: Introduction. – Models. – Thermodynamics. – Networks. – Networks for Transport Across Membranes. – Feedback Networks. – Stability. – References. – Subject Index.

What fundamental ideas and concepts can physics contribute to the analysis of complex systems such as those in biology and ecology? This book shows that thermodynamics – as used in physical systems analysis – has in the last ten years provided new concepts for the analysis of systems far from thermal equilibrium, and that these concepts can be used for describing and modelling biological systems as well. Although thermodynamics is the physical basis of the book, the language used is that of networks of bond graphs. A variety of examples is presented to demonstrate how this language is applied and how it leads to formulations of models for particular biological phenomena in such a way as to include the basic laws of thermodynamics. This new edition has been expanded by including a section on a network model for photoreception and additional examples of feedback networks for excitable systems.

Lasers in Photomedicine and Photobiology

Proceedings of the European Physical Society, Quantum Electronics Division, Conference, Florence, Italy, September 3–6, 1979

Editors: **R. Pratesi, C. A. Sacchi**

1980. 108 figures, 20 tables. XIII, 235 pages
(Springer Series in Optical Sciences, Volume 22)
ISBN 3-540-10178-0

Contents: General Introduction to Photomedicine and Photobiology. – Photodynamical Therapy of Tumors. – Photodermatology. – Phototherapy of Hyperbilirubinemia. – Absorption and Fluorescence Spectroscopy. – Raman and Picosecond Spectroscopy.

Physical and Biological Processing of Images

Proceedings of an International Symposium Organised by the Rank Prize Funds, London, England, 27–29 September, 1982

Editors: **O. J. Braddick, A. C. Sleigh**

1983. 227 figures. XI, 403 pages
(Springer Series in Information Sciences, Volume 11)
ISBN 3-540-12108-0

Contents: Overviews. – Local Spatial Operations on the Image. – Early Stages of Image Interpretation. – Pattern Recognition. – Spatially Analogue Processes. – Higher Level Representations in Image Processing. – Postscript. – Index of Contributors.

L. Reimer

Transmission Electron Microscopy

1983. Approx. 238 figures, approx. 26 tables. In preparation
(Springer Series in Optical Sciences, Volume 36)
ISBN 3-540-11794-6

Springer-Verlag Berlin Heidelberg New York Tokyo